Make Your Own

Wine

真正的釀酒，不是只有浸泡，更要完全發酵！本書將指導你如何運用輔助工具以及正確步驟，簡簡單單就能釀造出各式水果酒、乾果酒，甚至難得一見的氣泡酒！

釀一瓶自己的酒

氣泡酒、水果酒、乾果酒

釀酒專家 錢薇●著

朱雀文化

初學者的最佳指南

爲了防止食品腐敗，人們常以加熱、低溫、乾燥脫水、醃漬或發酵等各種加工技術加以保藏。「酒」就是一種發酵產品之一，因其原料狀態改變甚多，且需經長時間才得以釀成，故與味噌、醬油等均被特稱爲「釀造」食品，基本上，任何含有糖質的原料都可用來「釀酒」。

酒有可供作婚喪喜慶祭品、交際工具、壓力解消劑、增加飲食之樂趣、調製飲食或藥引及擔稅物資等效用，可說是一種民生必需品。「釀酒」雖然不難，世界各個國家、各個角落的居民，都會利用當地所產的不同原料釀造出所謂「酒」這個產品來，唯欲釀製好酒卻屬不易，必需將「技術」和「藝術」加以結合方能有成。

我國加入世界貿易組織，正式開放民間釀酒後，與酒相關的行業包括「酒廠」、「酒莊」與「酒類進口商」等迅速增加，許多民眾也以 DIY 的方式自釀自飲，唯許多民眾甚至業者對「釀酒技術」十分陌生，觀念也不正確；提供農民、業者「農產品加工與技術」之服務與輔導一向爲本廠設立宗旨之一，在此需求下開設許多釀酒研習課程，錢薇女士爲本廠特聘釀酒專家，協助本廠諸多學術科課程，她的專業技術與工作態度獲得學員們一致的讚賞。

錢女士畢業於淡江大學外文系，赴美後進入加州 Sacramento City College 社會系，畢業後因志趣而轉至 UC Davis 葡萄釀酒系研讀三年有餘，且成爲「美國葡萄及釀酒學會」及「美國葡萄釀酒教育協會」會員，此外她也熱心公益，爲美國加州 Sacramento 市華人工商協會創會會員，以及世界華人工商婦女企管協會會員之一。

欣聞錢女士將所學釀酒知識與技術彙集出版，詳讀本書即使是初學者也能自己釀出美酒，而對從事釀酒業的技術專家而言，這本書也有相當的參考價值。

柯文慶

國立中興大學食品科學系教授
兼農資學院食品加工廠廠長

Best Guide of Winemaking

The food industry uses many methods to preserve fresh products. Preservation techniques include heat, cold storage, dehydration, fermentation, marinating with sugar or salt, etc.

In the current economy, wine provides a significant share of taxable income for the government. Around the world, locally grown agricultural products are used as ingredients to make wine. Last year, Taiwan became a member of the World Trade Organization (WTO), allowing private companies and individuals to operate wineries. Since 2002, many wineries became established in Taiwan, but most operators lack the proper knowledge and techniques required for good winemaking. Many rely on local traditions to produce grape and fruit wines that are passed down from previous generations through word of mouth. It is not difficult to make wine, but producing a consistently high quality wine requires not only knowledge of scientific techniques but also understanding the art of winemaking.

The department of food sciences at National Chung Hsing University realized the importance of teaching winemaking techniques to support the booming winemaking industry. A series of grape and fruit winemaking courses was offered last year and proved highly successful with an attendance of over one thousand students.

The instructor, Theresa Lin was invited to teach the course, "The Basic Winemaking Process" at the food sciences department of National Chung Hsing University. With her professional credentials and in the field training, the course was very beneficial and highly received by her students.

Theresa is a graduate of Tamkung University in Taiwan. She also studied Social Sciences at Sacramento City College in California and Viticulture & Enology at the University of California, Davis. Theresa is a member of the "American Society of Viticulture & Enology", "American Winemaking Education Association" and is an active participant of numerous community service organizations.

With her extensive background, Theresa is the author of "home winemaking handbook" that provides basic techniques, equipment, ingredients and winemaking recipes. This comprehensive handbook is recommended for all audiences; it provides a good start for beginner winemakers and is a valuable guide for readers with prior winemaking experience.

Professor wen-ching Ko, Phd

Director Food Processing Plant of College of Agriculture
National Chung Hsing University

釀一瓶獨一無二的美酒

本書是爲了想要自己親手，以科學方法釀製美味葡萄酒及水果酒的人而寫的。

台灣民間釀製葡萄酒和水果酒已有很長久的傳統，口耳相傳的秘訣是以水果加上特定比例的糖，放置一旁，任水果自然發酵；時間到了，取出上層的澄清液，就完成了。這種釀酒方法可以說完全是靠運氣，依賴原料表面及空氣中含有的野生酵母菌，將水果中含有的糖分解，進行發酵作用，因而產生酒精；成功或失敗完全不能掌握，只能任憑天候變化，順其自然，無法決定成品是美味抑或酸敗。

民間行之有年的另外一種方法，是將水果或藥材浸泡在米酒或烈酒中，經過一段時間熟成。這種方法只是酒的加工，製出浸泡酒，也不能算是釀酒。

出版本書的目的，是希望將歐洲及美國各大酒莊在釀酒時，所採用的最先進最科學且專業的程序，以最簡單且易於操作的方法介紹給大家，讓讀者了解看似複雜難懂的專業釀酒，其實一點都不神秘。即便是價格昂貴的釀酒設備，也可以用最基本的工具取代。

書中分成三大章節：

〈基礎篇〉介紹專業釀酒的原理及基礎步驟。〈釀酒篇〉將台灣盛產的四季新鮮水果、常見的乾果，以及濃縮葡萄汁運用來釀製可口美酒，輔以詳細的步驟說明，讓讀者輕鬆學會釀酒。〈技術篇〉公開所有釀酒技術，以及操作技巧，讓每個人都也可以成爲釀酒高手。

釀酒是最能滿足自我成就的嗜好之一，只要懂得方法，利用簡單的工具，輕而易舉就可以釀出經濟實惠且美味的酒。看著自己釀造出來一瓶瓶有生命的美酒，聞著它，嘗著它，心中的喜悅迫不及待想要和人分享，不是嗎？

感謝好友Linda一直以來的鼓勵與協助，友誼就和美酒一樣，時間愈久，愈醇厚。

錢薇

聯絡方式
傳真：（02）23583078
e-mail：chienweilin2002@pchome.com.tw

Make Your Own Wine

This book is written for individuals who want to create delicious grape and fruit wines using the latest scientific methods. In Taiwan there is a long local tradition of making grape and other fruit wines. The secret, transmitted by word of mouth, is to combine the right proportion of fruit with sugar and let the fruit ferment naturally for a period of time. This method of winemaking depends totally on luck and relies on the wild yeast that exists on the fruit surface. The yeast and air break down sugars in the fruit to carry on fermentation, and produce alcohol. The success or failure of the project does not rest within human intervention and control; instead, success or failure is determined by the vagaries of climate and ingredients. Another method has also been used for quite some time by many locals. This process involves soaking the fruit or medicinal ingredients in rice wine or strong wine for a period of time, and then serving. This method is really only a type of processing that produces flavored wines and should not be regarded as true winemaking.

The purpose of this book is to introduce the simple, scientific, and professional methods used by winemakers in advanced winemaking areas in Europe and America. After reading this book readers will learn that winemaking, which seems complex, expensive and difficult to understand, is in fact not mysterious at all. For example, expensive winemaking equipment can be substituted with very basic tools.

This book is divided into three chapters:

"Basic" introduces the principals and basic procedures of professional winemaking. "Winemaking" introduces the four seasonal fresh fruits in Taiwan, commonly seen dried nuts, and delightful wines made from grape juice concentrate. The procedures are simple and clear, so that the readers can learn the art of winemaking easily and quickly. "Advanced Techniques" introduces all the open-to-the-public winemaking techniques. Each method is simple to learn, so that you too can quickly become an expert in the art of making wine.

I want to thank my best friend Linda for her encouragement and help. Friendship is just like a good wine: the longer it ages, the more enjoyable and pure it becomes.

Chienwei

c o n t e n t s

Chapter I Basic

基礎篇

Chapter II Winemaking

釀酒篇

Chapter III
Advanced Techniques

技術篇

附錄 Index

Basic

Chapter I

釀酒的基本原理 The Principles of Winemaking
釀酒的步驟 The Process of Winemaking
釀酒過程應注意的要點 Things to watch for during Winemaking
家庭釀酒基本設備 Basic Equipments for Winemaking
釀酒最常使用的原料 Common Ingredients for Winemaking
釀酒必要的添加物 Must Have Additives

1

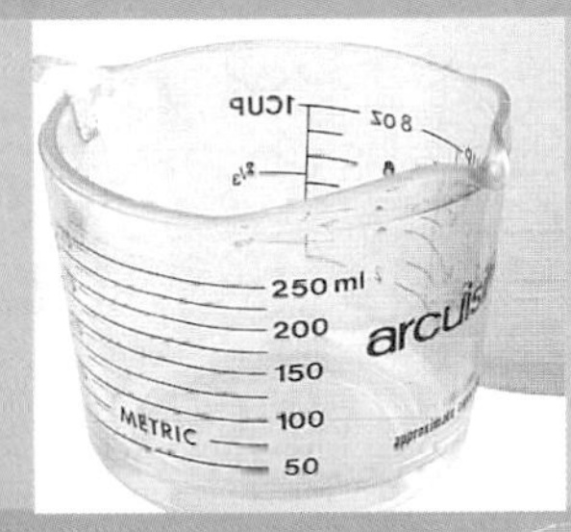

基礎篇

釀酒的基本原理 The Principles of Winemaking

釀酒的步驟 The Process of Winemaking

釀酒過程應注意的要點 Things to watch for during Winemaking

家庭釀酒基本設備 Basic Equipments for Winemaking

釀酒最常使用的原料 Common Ingredients for Winemaking

釀酒必要的添加物 Must Have Additives

釀酒的基本原理

酵母菌是一個單細胞植物，自由飄浮在空氣中或沾附在各種物體的表面，當酵母菌遇到了糖和空氣，這個單細胞植物立刻吸取糖的營養，呼吸空氣中的氧，進行生長分裂繁殖第二代的神聖任務，這個過程就是我們所謂的「發酵作用」。二氧化碳和酒精即為發酵作用的副產品──釀酒的原理就是如此簡單。

釀出好酒的訣竅，主要就是學習使用正確的酵母菌，將水果中的糖，在理想的環境中，發酵轉化成酒精的方法。

酵母菌在22％比重的糖液中，生長分裂最旺盛。所以，釀酒的首要條件就是：**將果汁調節到酵母菌最喜愛的環境，添加砂糖或濃縮果汁可以達到這個目的**。酵母菌還需要礦物質和維生素B群等營養素，來幫助生長、維持生命；所以第一次發酵時，在果汁中添加適量的酵母營養劑，可以促進酵母的活性，快速分裂繁殖。外界的溫度也影響酵母的生長，15～26℃最理想，釀酒時要盡量將果汁或酒醪保存在這個溫度範圍內。

有了上述的理想環境，酵母菌將能很快樂地進行優質的發酵作用。此時還要保護它不受到其它壞菌和雜菌的侵擾、感染，因為空氣中充滿了成千上萬種野生酵母菌及霉菌。野生酵母菌大部份是雜菌，不適合釀酒，就算有少部分能釀酒，但品質不穩定，很難每次都能釀出好喝的酒。

釀酒專用的酵母菌，是經過馴化、篩選、培育，而成為純化的菌株，它的唯一用途就是釀酒。為了避免釀酒時受到野生菌的感染，一定要將使用的器具清洗乾淨並殺菌，所有的容器都要加蓋，減少與空氣的接觸。

酵母菌發酵作用，只是釀酒過程中的一個步驟而已。不只葡萄，幾乎所有的水果都能釀酒，甚至蔬菜、花卉也可以釀酒。台灣每年4、5月盛產的熟黃梅就是釀酒的上好原料，但是它的酸度相當高，若不經過降酸處理，釀好的酒既酸又澀，難以下嚥。另一個例子是火龍果，胭紅的果汁可以釀出色彩鮮艷的酒，但口感平淡，缺乏風味，必須先將火龍果汁做增酸處理。因此，糖度和酸度的平衡，也是釀酒的基本條件。

當發酵作用完成後，酒本身仍在進行各種不同的變化，果皮果肉等雜質逐漸沈澱，酵母菌細胞分裂時產生的蛋白質，以及各種酯、醇、酚等化合物，互相串連結合，營造出濃郁的酒香，這也是新酒釀好後，必須要進行「熟成」的原因。陳年熟成的時間沒有固定的標準，由釀酒師自己設定。

釀酒的大前題是盡量將水果的色、香、味充份表現到酒裡，讓水果展現出新的風華！

The Principles of Winemaking

The basic principles of winemaking are actually quite simple. The key to making good wine is to learn how to create the ideal conditions for yeast to break down the sugars within the fruit to produce alcohol.

Yeast is a single-celled organism that is found floating freely in air or on the surface of many objects. When yeast meets sugar and air, it immediately absorbs nutrients from the sugar and oxygen from the air, and then carries out the processes of breaking down the sugar in pursuit of its sacred mission of reproduction. This process is referred to as fermentation. Carbon dioxide and alcohol are the by products of fermentation.

The most efficient proportion of yeast to sugar liquid, under which it grows and breaks down the sugar most vigorously, is 22 percent. Therefore, the basic condition of good winemaking is **to adjust the sugar content of the juice to the most suitable to the yeast, by adding granulated sugar or juice concentrate**.

Yeast also requires nutrients, such as minerals and vitamin B, to facilitate growth and to maintain life. Thus,

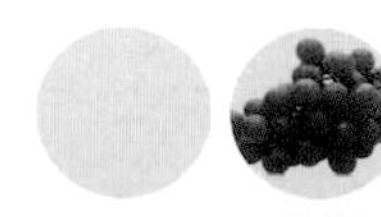

when primary fermentation commences, add a suitable amount of yeast nutrients to promote vigorous yeast growth and to accelerate the breaking down of sugars and the propagation of the yeast.

The outside temperature also affects the growth of the yeast. Ideally, temperatures between 15 ~ 26˚C are the best for yeast to breed. Yeast will efficiently ferment sugar into alcohol under the ideal conditions. At the same time, it is important to protect the yeast from invasion and infection by unwanted and undesirable bacteria and other micro-organisms. There are in fact hundreds of strains of wild yeast and bacteria. Wild yeast strains are not recommended for making quality wine. Small amounts may be used, but its varying quality makes it difficult to produce good wine every time.

The yeast strains used in winemaking today have been domesticated, selected, and cultivated for purity and reliability. In order to prevent contamination by wild yeasts, cleanliness and good hygiene must be maintained throughout each stage of the winemaking process. All equipment must be sanitized and containers should be covered to reduce contact with the air.

Not just grapes, but almost any fruit, may be fermented to produce alcohol. Even vegetables or flowers can be used for making wine. Another fundamental attribute to successful winemaking is to balance the sweetness and sourness of the fruit. In Taiwan, seasonal yellow plums that ripen between April and May are ideal ingredients for winemaking, but the degree of sourness is high. If the sour flavor is not reduced, the wine will be sour, dry, and unpalatable.

Almost every bottle of wine improves with time. When primary fermentation is complete, the wine is still progressing through different levels of changes. The must gradually becomes sediment, while the cells of the yeast break down and produce chemicals such as proteins, esters, ethyls, and polyphenols. All come together to yield an aroma and condition the wine. This is the reason wine needs to be aged. There is no standard time limit for aging. It is generally decided by the winemaker.

The art of winemaking is to leverage the color, aroma and flavor of the fruit to create wine and let the fruit present a new, more elegant face.

釀酒的步驟

這一章將說明釀酒的7個步驟，讓你了解每一個過程大約需要多少時間。

工作過程／需要的時間

1.將所有器具、容器清洗乾淨、消毒殺菌。（約40分鐘）

2.原料準備齊全，清洗水果，揀除腐壞或未成熟的水果；按照酒譜，將果粒切碎、壓汁、搗爛、煮軟，製成酒醪。（約2小時）

3.將準備好的酒醪，放入發酵桶，按照酒譜說明，逐項加入所需的原料及酵母，蓋好發酵桶，放在23°C左右的陰暗處，每天攪拌兩次。（約5～10天，依溫度及酵母品種而不同）

4.虹吸或過濾，將發酵桶之酒醪，丟棄果皮殘渣，換入第二次發酵之果酒瓶，用橡皮塞封口，加入發酵鎖，直到發酵作用停止。（約10天）

5.虹吸換瓶，加入澄清劑。（約10天）

6.虹吸換瓶，加入二氧化硫，靜置於溫度20°C以下，避免光線。（水果酒1～3個月/葡萄酒3～6個月）

7.加入安定劑，虹吸裝瓶；裝瓶後，按食譜說明，繼續熟成。（約3～9個月）

※酒醪這個專有名詞，指的是釀酒原料經過切碎、壓汁、搗爛、煮軟等程序，準備成適合釀酒的狀態。

The Process of Winemaking

This chapter describes 7 steps in the winemaking process with an estimate of the amount of time required for each step.

The Winemaking Process/Time Required

1. Clean and sterilize equipment and bottles. (About 40 Minutes)
2. Prepare all the ingredients and clean the fruit. Discard all decayed, rotten or unripened fruit. Follow the recipe and crush the diced fruit, then squeeze into juice. Stir well until mashed and boil until softened to form the must. (About 2 hours)
3. Transfer the must to the fermentor. Follow the steps in the recipe, gradually adding the ingredients and yeast. Cover and move to a cool and dark location that maintains a temperature of 23℃. Uncover and stir twice a day.This step takes approximately 10 days, but time may vary with the temperature and strain of yeast, so adjust as necessary.
4. Siphon or strain, discarding any solid fruit particulates from the must. Transfer to bottles to begin secondary fermentation. Seal with rubber bung sulphite to create an airtight cover until fermentation stops.(About 10 days)
5. Siphon and transfer to bottles, add clearing agent. (10 days)
6. Siphon and transfer to bottles, add sulfite. Move the bottles to an environment that is under 20℃, no light is allowed. (1~ 3 months for fruit wine. 3 ~ 6 months for grape wine)
7. Add stabilizer, siphon and bottle. Follow the recipe and continue to age. (About 3 ~ 9 months)

釀酒過程應注意的要點

1. 整體的清潔及消毒。
2. 水果成熟度要夠，新鮮、無腐爛果實。
3. 盡量避免和空氣接觸（第一次發酵時例外）。
4. 溫度控制適當，水果酒為了保存果香及風味，發酵時溫度不要超過20℃，白葡萄酒和水果酒相同，紅葡萄酒可以達到25～30℃。
5. 儲存時要放在陰暗無光線照射之處。
6. 裝瓶時除了酒瓶要清潔殺菌，周遭環境也要保持清潔，不能有蒼蠅及風沙。

Things to Watch for during Winemaking

1. Be sure to maintain cleanliness and sterilization throughout every stage of the winemaking process.
2. The fruits should be ripe and fresh without any spoilage.
3. Prevent contact with the air as much as possible (excludes the first fermentation).
4. Control the temperature well. In order to preserve the fruit's flavor and aroma, the fermentation temperature should not exceed 20℃ for white or fruit wines. Red wine can be permitted to reach 25℃ to 30℃.
5. Store the bottles in cool and dark location where sunlight does not reach.
6. Clean and sterilize the bottles thoroughly before bottling. Maintain the cleanliness of the working area. Do not permit insects or dust to contaminate the area.

家庭釀酒基本設備

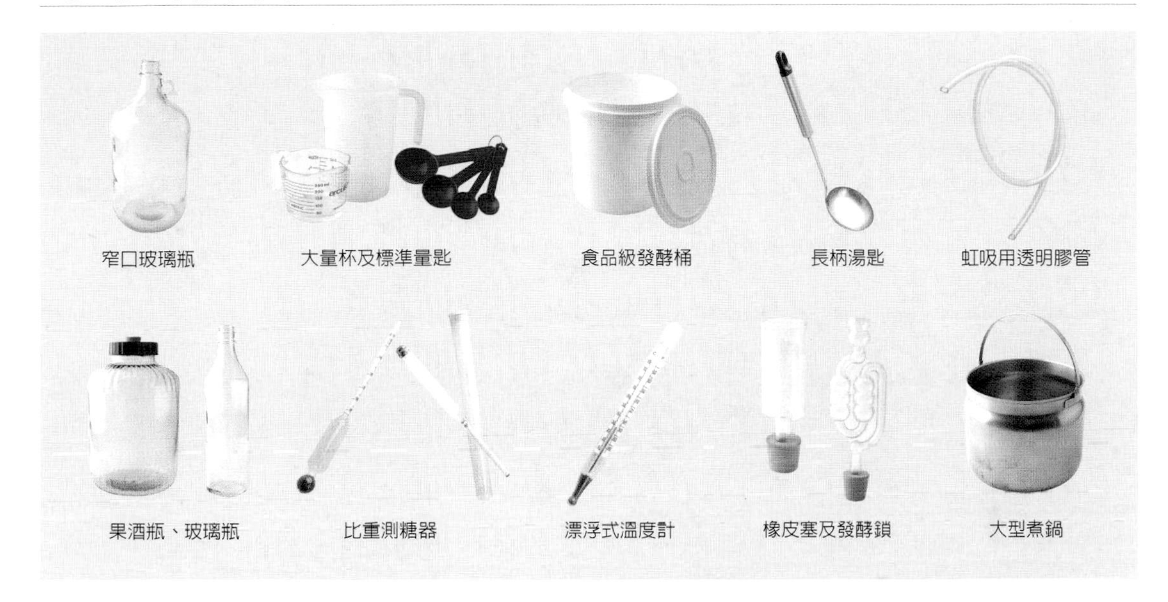

窄口玻璃瓶　大量杯及標準量匙　食品級發酵桶　長柄湯匙　虹吸用透明膠管

果酒瓶、玻璃瓶　比重測糖器　漂浮式溫度計　橡皮塞及發酵鎖　大型煮鍋

1.窄口玻璃瓶：用來盛裝4000c.c.（4公升）殺菌藥水，釀酒前所有的器具都要清洗、殺菌，高溫高壓的熱水是最好的殺菌方法，家庭釀酒沒有高壓的熱水，則用煮開的熱水即可。也可自製殺菌藥水：將60公克二氧化硫粉末，溶入4公升冷水，裝在玻璃瓶中使用，可在1個月期限內使用。

2.大量杯及標準量匙：釀酒和做科學實驗相同，份量絕對不可馬虎。

3.食品級塑膠桶：大部份酒譜在第一次發酵時，水果及果汁的總容量約在25～30公升之間；第一次發酵桶要有50公升的容量，發酵時才不會滿溢。切記，不要買五顏六色的塑膠桶，因為酒醪會萃取染料中的重金屬。新桶使用前，裝滿溫水，加入2湯匙蘇打粉，浸泡過夜，再用清水刷洗乾淨。

4.長柄湯匙：不銹鋼、塑膠或木製材質，不要用鋁製品來攪拌水果原料及添加物。

5.虹吸用透明膠管：約6公尺長、小姆指粗細的透明膠管，可減少空氣和酒醪的接觸。

6.果酒瓶2個及小型玻璃瓶數個：第二次發酵時，果酒瓶一定要裝滿，可準備18公升容量的果酒瓶2個及1公升玻璃瓶2～3個。

7.比重測糖器：釀酒必備的工具之一，利用比重的原理，測量出果汁及酒醪的含糖量。

8.漂浮式溫度計：放在發酵桶裡，可以隨時知道發酵中酒醪的溫度，溫度過高酵母會死掉，溫度太低酵母則會進入冬眠，發酵作用也會停止。

9.橡皮塞及發酵鎖：為家庭釀酒特製的工具。第二次發酵時，果酒瓶中因發酵作用產生的二氧化碳，可經由發酵鎖排出桶外，且阻止外面的空氣進入。當發酵鎖不再有氣泡湧出時，就可以知道發酵作用已經停止。

10.塑膠布：若第一次發酵桶沒有蓋子的話，發酵時要用塑膠布將桶口封好，但不要密封，讓空氣可以進入，但要避免蒼蠅或灰塵進入發酵桶。

11.筆記本：詳細記錄釀酒的每一個過程及細節，好作為自己的參考。

12.大型煮鍋：1個。

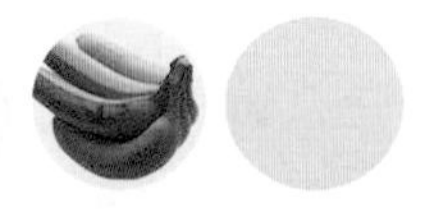

Basic Equipment for Winemaking

1. Glass Carboys: Use to hold up to 4,000c.c. (4 liters) sanitizing solution. All equipment must be thoroughly cleaned and sterilized before use. High temperature and high pressure hot water is a very good way to sterilize. Home winemakers usually do not have access to pressurized water, so boiling hot water will suffice. Another option to sanitize the equipments at home is by dissolving 60 grams of sulfite powder in 4 liters of cold water. Transfer to glass carboy. This solution maintains a shelf life of 1 month.

2. Large measuring cups and standard measuring spoons: During winemaking, precise measurements are crucial.

3. Food-grade plastic bucket: The total content of fruit and juice should be about 25 ~ 30 liters during the first fermentation. A 50-liter pail is required during the first fermentation, so that the solution will not overflow. Remember, do not purchase colored plastic pails because the must will absorb the heavy metals from the dyes, giving it a metallic flavor. Fill the new pails with warm water, add 2 spoonfuls of baking soda and soak overnight, then brush and rinse out with water.

4. Long Stirring Paddles: Stainless steel, plastic or wooden. Do not use aluminum paddles to stir the fruit must and other ingredients.

5. Transparent Tube for Siphoning: 6 meter long transparent tubes of approximately a pinky's width, used to reduce contact between the must and the air.

6. 2 large glass bottles and a couple of small glass bottles: During the secondary fermentation, the large glass bottles should be completely filled. Prepare 2 18-liter glass bottles and 2 to 3 1-liter glass bottles.

7. Hydrometer: This is one of the most important pieces of equipment for making wine. It is used to measure the relative specific gravity of the sugar in juice or the must.

8. Floating Thermometer: Use in fermentor to measure the temperature of the fermenting must. The yeast will die if the temperature too high, or go dormant if too low; either may cause fermentation to stop.

9. Rubber Bungs/Stoppers and Airlock: Two special tools for winemaking. During the secondary fermentation, carbon dioxide produced in the wine bottles is vented through the airlock, which prevents air from entering the bottles. When no more foam comes out of the airlock, it is a sign that fermentation has ended.

10. Plastic Wrap: During the primary fermentation, if the fermentor does not have a cover, loosely seal the opening with the plastic so that air can enter, but insects and dust are blocked.

11. Notebooks: Make notes about each step and capture details to use as reference for the next batch.

12. A large pot

釀酒最常使用的原料

「Wine」這個字指的是用葡萄釀製而成的酒，是葡萄酒的專用字；若是用其它原料釀製而成的酒，必需在Wine的前面加上所使用的原料，例如：「Peach Wine」桃子酒，「Cherry Wine」櫻桃酒等。

大自然賜予的水果、鮮花、蔬菜，以及五穀都是釀酒的好材料，我們可以購買專為家庭釀酒而設計出的小包裝釀酒酵母，使得家庭釀酒更加輕而易舉。

下面列舉出釀酒最常使用的原料：

1.新鮮水果

2.新鮮蔬菜

3.乾果：葡萄、藍莓、無花果、香蕉、鳳梨等各種果乾
4.新鮮葡萄汁或水果汁
5.濃縮葡萄汁及各種水果汁
6.可食用的鮮花

乾果

果汁

Common Ingredients for Winemaking

"Wine" generally refers to the products of grapes and is considered a special term for grape-based alcohols. Alcohols made from other ingredients should have the ingredient's name in front of "wine," for example, peach wine or cherry wine.

Nature provides a wide selection of fruits, flowers, vegetables and grains that can be used for making wine. With improvements in biochemical technology, highly-developed yeasts are available at select stores making home winemaking even easier to learn.

The following are the commonly used ingredients in winemaking:

1. Fresh fruit
2. Fresh vegetables
3. Dried fruit: grapes, blueberries, figs, bananas, pineapple, etc.
4. Fresh grape juice or other fruit juice
5. Grape juice and other fruit juice concentrate.
6. Edible flowers

釀酒必要的添加物

1.酵母：幾乎已經沒有酒莊在釀酒時，是依靠水果表面的野生酵母釀酒，因為可以購買各種不同品牌的，專門為釀酒而培育的純種酵母。理想的酵母能夠適應不同的溫度變化及酒精濃度，發酵時不會產生不好的氣味及副產品；發酵結束時，會自動沈澱在底部，使上面的酒液澄清。

2.糖：釀酒時，水果中的含糖量，可決定釀出來的酒含有多少酒精度，奇妙吧！只要記得水果的含糖量除以2，就是酒精度；為了能釀出酒精度10～14%的葡萄酒，我們就添加蔗糖來提高果汁中的糖到20～28度。果糖及蜂蜜也是很好的代用品。葡萄糖在發酵時無法完全分解，會增加酒的質感；但不能取代蔗糖，只能加少量在果汁中。

3.單寧：葡萄皮、莖、種子中含有大量的單寧，紅酒因含有單寧而醞釀出它獨特的風味及耐久存的特性。熱帶水果比較少含有單寧，所以使用熱帶水果釀酒時，要適量的添加一點釀酒專用的單寧，增加水果酒的風味及複雜性。

4.水果酸：熱帶水果比較偏甜，例如芒果、荔枝等；而釀酒時，甜和酸必須平衡，才能釀出口感濃郁的酒。只有甜味的果汁，就要增加酸來提高果汁的酸度，最好的酸是酒石酸、蘋果酸和檸檬酸的混合酸。將這三種酸按不同的比例混合，添加到果汁中，才能達到完美的效果。

5.果膠酵素：所有的水果中都含有果膠，壓汁時不易將果汁充份搾取，而且會使搾出的汁液混濁；加入果膠酵素，可以幫助果膠分解，汁液澄清，同時增加出汁量。

6.二氧化硫：果汁發酵時會自然產生二氧化硫，以防止果汁被雜菌污染及氧化；但自然產生的二氧化硫量非常少，因此酒莊釀酒時，會在不同的程序中添加二氧化硫增強它的功效。二氧化硫在釀酒中具有不可取代的重要性，但因為添加量極少，除了對二氧化硫敏感的人以外，對健康不會有任何影響。

7.酵母營養劑：酵母菌是單細胞植物，需要維生物、礦物質維持它的生命及生長、繁殖，酒廠在釀酒時都添加少量的酵母菌營養劑，幫助酵母菌旺盛地進行發酵作用。

8.澄清劑：它的作用是將已發酵完成的酒液中所含有的固體懸浮物，經由澄清劑的吸附作用，彼此凝結而沈澱，使酒液澄清。包括皀土及吉利丁粉都屬澄清劑。

9.橡木：橡木桶儲酒可以增加酒的風味及質感，但要適可而止，太濃的橡木風味，會蓋過酒本身的香醇，如果希望酒帶有清爽的果香，就不要使用橡木桶儲存。

10.安定劑：酒譜中添加的安定劑可增加裝瓶酒的穩定性。專業的釀酒廠有很精密的過濾設備，在裝瓶前可以將酒液中殘存的酵母菌和微生物過濾清除。而家庭釀酒在裝瓶前，就得加入安定劑來抑止酒液中殘存的酵母菌和微生物，以避免裝瓶後還產生發酵或感染現象，維持瓶中酒的穩定性。安定劑亦有提高酒中甜味的功效。

Must Have Additives

1. Yeast: winemakers rarely use the wild yeasts that occur naturally on the surface of fruit. There are many strains of pure yeast cultivated for winemaking that are commercially available. The ideal yeast can adjust to the different temperature changes as well as the level of alcohol and will not produce unwanted flavors or products during fermentation. After fermentation, it will become sediment and sink to the bottom, leaving the clear wine on top.
2. Sugar: The sugar gravity determines the alcohol gravity. Wonderful, right? You only need to remember that the sugar gravity in fruit is twice the alcohol gravity. To produce grape wine of 10 ~ 14% alcohol, sugar must first be added to increase the sugar gravity to 20 ~ 28%. The grape varieties grown in Taiwan used for wine-making are generally the black queen red or the golden muscat white. These two varieties of grapes contain approximately 14 ~ 18 grams of fructose. Fructose and honey are also good substitutes for sugar. Glucose does not dissolve completely in fermentation and increases the body of the wine, though it can not take the place of sugar. Add only small amounts to the juice.
3. Tannin: Grape peels, stems and seeds all contain large quantities of tannins. Red wine derives its characteristic flavor from the tannins, which also have preservative effects. Tropical fruits seldom contain tannins, so be sure to add a suitable amount of tannins to enhance the flavor and body of the wine when using tropical fruit.
4. Fruit Acid: Tropical fruits such as mangoes and lychees are generally sweeter. It is important to balance the sweet and sour levels to create a flavorful wine. Sweet juice needs acid to increase the sourness of the juice. The best souring agent is a combination acid of tartaric acid, malic acid and citric acid. Combine these three acids in the right proportion and add to the juice to create the right taste.
5. Pectin Enzymes: All fruits contain pectin. It is difficult to squeeze the juice out of the fruit completely, and the juice is often cloudy due to pectin. Adding pectin enzymes will help break down pectin, clarify the juice and increase the amount of liquid.
6. Sulphite: When juice ferments, it produces carbon dioxide naturally. In order to prevent the juice from contamination and oxidization, sulfur dioxide is added to increase the fermentation and to supplement the reduced amounts of carbon dioxide in the juice.
   Sulfur dioxide plays an important role during winemaking and there is not a substitute. The amount of sulfur dioxide added is minimal so it has no effect on the health except for those who are allergic to sulfur dioxide.
7. Yeast nutrients: Yeast is single celled organism that requires nutrients and minerals to live, grow and multiply. Small amounts of yeast nutrients are added during the wine making process to help the yeast carry out fermentation.
8. Clearing Agent: The function of the clearing agent is to absorb the floating particulates after fermentation, coagulation and sedimentation, so as to clarify the wine. Bentonite or gelatin powder are both used as clearing agents.
9. Oak: Oak barrels and pails enhance the flavor and quality of the wine, but be careful not to overuse. An overly heavy oak flavor will mask the aroma of the wine. Do not use oak barrels if your intent is to produce wine with a clear, light, and fruity aroma.
10. Stabilizer: Adding stabilizers will increase the stability of bottled wine. Professional wineries use extremely precise filtration equipment that removes all the leftover yeast and microorganisms before bottling. In home winemaking, before bottling, stabilizers must be added to terminate the leftover yeast and microorganisms in the wine and to prevent fermentation and contamination after bottling. Stabilizers can also increase the level of the sweetness in the wine.

Winemak

Chapter II

台灣特產水果酒 Taiwan Specialty Fruit Wines
四季新鮮水果酒 Four-Season Fresh Fruit Wine
蜂蜜酒 Mead
乾果釀酒 Dry Fruit & Nut Wine
濃縮葡萄汁釀酒 Grape Juice Concentrate Wines

釀酒篇

台灣特產水果酒 Taiwan Specialty Fruit Wines
四季新鮮水果酒 Four-Season Fresh Fruit Wine
蜂蜜酒 Mead
乾果釀酒 Dry Fruit & Nut Wine
濃縮葡萄汁釀酒 Grape Juice Concentrate Wines

Winemaking

總論

這一篇提供許多利用各種不同水果、常見乾果，以及濃縮果汁為原料的釀酒方式。台灣天氣比較熱，在家裡最好能找到一處陰暗的地方，若有地下室最為理想；因為果汁發酵時，若溫度超過25°C，水果的香味就會揮發而失去了果香，釀好的酒就缺少了水果酒的香醇風味。

我稱呼這些不同的釀酒方法為「酒譜」，每一道酒譜都是經過仔細計算份量，並反覆試驗過的。只要按部就班照著酒譜製作，並徹底做好清洗器具殺菌的工作，3個月到半年後，就能呼朋喚友來享受自製佳釀了。

酒譜上特別注明第一次發酵及第二次發酵兩個過程，第一次發酵時，需要空氣，幫助酵母呼吸、生長繁殖。發酵桶在第一次發酵時不能密閉。第二次發酵時，就要避免空氣和酒醪的接觸，每次虹吸換瓶時，要將果酒瓶裝滿，再加橡皮塞和發酵鎖。這裡有一個避免氧化的方法，準備一個18公升的果酒瓶做為第二次發酵的果酒瓶，以及兩、三個1公升的玻璃瓶；第一次發酵完成的酒醪先裝滿18公升的果酒瓶，多餘的裝入1公升的玻璃瓶裡，全部加橡皮塞及發酵鎖密封，下一次虹吸換瓶時，18公升的果酒瓶去除沉澱，酒醪已經裝不滿18公升的果酒瓶，這時用1公升裝的酒醪填滿，就不必擔心會有裝不滿的果酒瓶。

※酒譜中添加的安定劑，除了增加裝瓶酒的穩定性，亦有提高酒中甜味的功效。

※酒譜中添加的精製食用甘油，作用是增加酒的質感。

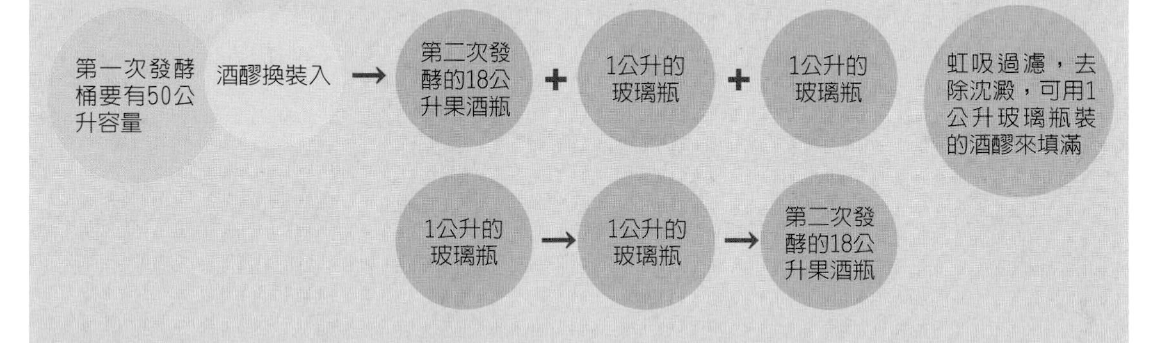

Conclusion

This chapter offers information on using different kinds of fruit, common dried fruits, and juice concentrate as the primary ingredients for making wines at home. As the weather in Taiwan is warmer, it is best to find a cool and dark location in the house to store the wine. A cellar is ideal, because when the juice ferments, if the temperature exceeds 25C, the flavor will evaporate and lose its fruit aroma.

In this book, I refer to my different methods of making wines as recipes. The portions of each ingredient in each wine recipe has been well calculated and repeatedly tested. If you follow the step-by-step instructions and keep your equipment clean and sanitized, In three to six months your family and friends can enjoy a bottle of homemade wine.

Each wine recipe clearly notes the processes of primary fermentation and secondary fermentation. During primary fermentation, air is required to help the yeast breathe, grow and multiply. The fermentor should not be sealed at this time. During secondary fermentation, prevent air from coming into contact with the must. Each time siphoning and bottling are performed, add rubber bungs and airlocks. To prevent oxidization, prepare an 18-liter glass bottle as the bottle for the second fermentation as well as two to three 1-liter glass bottles. First fill the 18-liter bottle with the primary fermented must, then fill the remaining 1-liter glass bottles with the remaining of the must. Seal with rubber bungs and airlocks. When next siphoning comes, the 18-liter bottle will have space after the sediment is removed. Fill the bottle with the must from the 1-liter bottles to fill-up the empty space.

The fermentor should be able to hold 50 liters during the primary fermentation.

Transfer the must to an18-liter glass bottle, several1-liter glass bottle for the second fermentationSiphon and strain out the sediment. Use the must from the 1-liter glass bottle to refillthe 18-liter glass bottle during the secondary fermentation.

台灣特產水果酒

亞熱帶台灣自古就是水果天堂，盛產時節常有生產過剩、果農賤售的情形；辛勤整年的果農如果學會了釀酒技術，就不會眼睜睜看著水果腐爛、血本無歸了。自己在家把多餘的水果釀成美味可口的水果酒，越陳越香醇，自用送禮皆相宜。

Taiwan Specialty Fruit Wines

Since subtropical Taiwan is a paradise of fruit, when fruits are in season, farmers are forced to sell a surplus of fruits at low prices. By leveraging basic winemaking techniques, these fruits could be bottled into delicious wines and share with friends.

Banana Wine
香蕉酒

香蕉酒濃郁香醇，喝時加冰塊及一片檸檬，或是對一點汽水，是夏日消暑的最佳雞尾酒。

Banana wine is thick and fragrant. Serve in a glass with ice cubes and a lemon wedge, or add a little soda. It is a great cocktail for fighting the summer heat.

■ 製成份量：約20公升
■ 完成時間：5個月可裝瓶

材料1：（第一次發酵）

黃熟香蕉	5.5公斤
糖	1.5公斤
熱開水	6公升
增酸劑	8小匙
柳橙	4粒
酵母營養劑	2小匙
單寧	2小匙
二氧化硫片（壓碎）	6片
果膠酵素	2小匙
冷水	9公升
水果酒酵母	1包（5公克）
溫水	1杯

材料2：（過程中添加）

糖	450公克
皂土	2公克
安定劑	300c.c.
二氧化硫	3片

Ingredients 1（The primary fermentation）

5.5 kgs. ripe yellow bananas
1.5 kgs. sugar
6 liters boiling water
8 tsp. acid blend
4 oranges
2 tsp. yeast nutrients
2 tsp. tannin
6 campden tablets（crushed）
2 tsp. pectin enzyme
8 liters cold water
1 pack champagne yeast（5 grams）
1 cup lukewarm water

Ingredients 2（Add during the process）

450 grams high gravity alcohol yeast
2 grams bentonite
300c.c. stabilizer
3 campden tablets

■ Finished Amount : about 20 liters
■ Finishing Time : 5 months before bottling

香蕉

柳橙汁

二氧化硫

糖

單寧粉

果膠酵素

❶ 香蕉切片加入開水和糖
❷ 香蕉切片煮軟
❸ 用取酒器取酒醪測糖度

做法

第一次發酵

1. 香蕉連皮切碎，放入大鍋裡加入4公升熱水，以小火煮30分鐘。
2. 將煮過的香蕉汁，以漏網過濾，倒入發酵桶；趁熱加入糖、增酸劑和2公升熱水。
3. 用力攪拌，直到糖充份溶解。
4. 4個柳橙擠汁、加入。
5. 依序加入酵母營養劑、單寧、二氧化硫片、果膠酵素及冷水攪拌均勻。
6. 用發酵桶蓋或塑膠布將發酵桶蓋好，放置1個小時。
7. 以溫度計測量，不要超過23℃。
8. 酵母加入1杯溫水（和體溫相同，不要超過38℃）攪拌，放置10～30分鐘，觀察整杯水已呈混濁，甚至有氣泡產生，此時倒入發酵桶，用力攪拌。
9. 將發酵桶加蓋、或用塑膠布蓋好，但不能密封，放置在23℃的暗處。
10. 待24小時後檢查，見泡沫浮在表面，發酵作用已經開始；若沒有發酵現象，再等一天，若仍沒有動靜，請參考P.111（釀酒Q&A）問題1的原因解答。
11. 每天用長柄匙，早晚各攪拌一次。

Steps

The primary fermentation

1. Crush the bananas finely with the skin on, then add to a pot with 4 liters of boiling water. Cook over heat for 30 minutes.
2. Pour the banana solution through a strainer to remove solid particles and transfer the must to a fermentor. Add sugar and acid blend as well as two liters of hot water.
3. Stir vigorously until the sugar dissolves completely.
4. Squeeze the four oranges and add the juice to step 3.
5. Add yeast nutrients, tannin, campden tablets, pectin enzyme and cold water in the listed order. Stir until well mixed.
6. Cover the fermentor with a lid or with a plastic wrap. Let sit for 1 hour.
7. Test with a thermometer; be sure not to let the temperature exceed 23℃.
8. Dissolve yeast in 1 cup of warm water (lukewarm, do not exceed 38℃). Stir well and let sit for 10 to 30 minutes until the yeast solution is cloudy, and foam forms on the surface. Add to the fermentor and stir vigorously to mix.

9. Cover the fermentor with a lid or with plastic wrap. Do not seal too tightly. Transfer the fermentor to a cool, dark corner with a temperature of 23°C.

10. Let sit for 24 hours, then open to examine. If foam has formed on the surface, fermentation has started. If not, wait for another day and check again. If there is still no foam on the surface, refer to Question & Answer 1 on p.111〔Wine Making Q&A〕.

11. Stir twice a day, once in the morning and again at night, with a long stirring paddle.

第二次發酵

12. 每隔1天，取一些酒醪，用比重測糖器檢測糖度，當含糖量降到5度以下時，用細網或虹吸過濾到第二次發酵用的果酒瓶裡，取出1,000c.c.酒醪加入450公克糖，溶解後倒回果酒瓶，用橡皮塞封口加上發酵鎖。放在陰暗處，溫度不要超過20°C。

13. 靜置1個月，虹吸換瓶，加3片二氧化硫片（先壓碎，以少許冷水溶解）。

14. 靜置15天，虹吸換瓶，加入2公克皂土澄清。

15. 10天後再次虹吸換瓶，加入安定劑300c.c.，靜置熟成3個月。

16. 裝瓶，在瓶內熟成5個月。

The secondary fermentation

12. Test a sample of the must with a hydrometer every other day to measure the gravity of sugar. When the sugar gravity is below 5, use a fine strainer or siphon to transfer the must into bottles to begin the secondary fermentation. With 450 grams of sugar in1,000c.c. of the must, let the sugar dissolve completely, then return to the fruit bottles. Seal with rubber bungs and airlocks. Put the bottles in a dark, cool corner where the temperature does not exceed 20°C.

13. Let sit for 1 month, siphon and rack to bottles, along with 3 pieces of campden tablets added (crushed first, then dissolved in little cold water). Racking is the process of siphoning wine into another container and leaving the sediments behind.

14. Let sit for 15 days, siphon and rack to bottles, add 2 grams of bentonite to clarify.

15. Let sit for another 10 days, then siphon and rack to bottles along with 300c.c. of stabilizer added. Let age for 3 months.

16. Bottle and age for 5 months.

Mango Wine

芒果酒

■ 製成份量：約25公升
■ 完成時間：3個月可裝瓶

加州大學釀酒系特別推薦芒果酒，用白葡萄酒的方法釀製，果香撲鼻、顏色金黃，好看又好喝。

Mango wine is highly recommended by the winemaking department of California University. The resulting wine is golden yellow, has a fruity aroma, with a pleasing appearance and excellent taste.

材料1：（第一次發酵）

芒果	8公斤
糖	4公斤
白葡萄濃縮汁	1公斤
(可以新鮮青葡萄去皮榨汁取代)	
增酸劑	12小匙
熱開水	6公升
酵母營養劑	2小匙
果膠酵素	2小匙
單寧	1小匙
二氧化硫片（壓碎）	5片
冷水	9公升
香檳酒酵母	1包（5公克）
溫水	1杯

材料2：（過程中添加）

皂土	2公克
二氧化硫片	6片
安定劑	300c.c.

Ingredients 1 (The primary fermentation)

8 kgs. mangoes
4 kgs. sugar
1 kgs. green grapes concentrate
12 tsp. acid blend
6 liters hot water
2 tsp. yeast nutrients
2 tsp. pectin enzyme
1 tsp. tannin
5 campden tablets(crushed)
9 liters cold water
1 pack(5 grams)champagne yeast
1 cup lukewarm water

Ingredients 2 (Add during the process)

2 grams bentonite
6 tablets sulfur dioxide
300c.c. stabilizer

■ Finished Amount : about 25 liters
■ Finishing Time : 3 months before bottling

芒果

二氧化硫

白葡萄濃縮汁

香檳酒酵母

糖

單寧粉

果膠酵素

❶ 芒果切片
❷ 添加酵母營養劑
❸ 用測糖器檢測糖度

做法

第一次發酵

1. 芒果去皮取果肉，與白葡萄濃縮汁、糖、增酸劑、熱開水，全部倒入第一次發酵桶裡，攪拌直到糖完全溶解。
2. 加入酵母營養劑、果膠酵素、單寧、二氧化硫片及冷水，攪拌均勻，蓋好發酵桶，靜置1個小時。
3. 酵母加入1杯溫水（38℃），放置10～30分鐘，加入發酵桶內，攪拌。
4. 將發酵桶加蓋、或用塑膠布蓋好，但不能密封，放置在23℃左右陰暗處。
5. 待24小時後檢查，見果肉浮在表面，表示發酵作用已經開始；若沒有發酵現象，再等24小時，若仍沒有動靜，請參考P.111（釀酒Q&A）問題1的原因解答。
6. 每天用長柄匙，早晚各攪拌一次。

Steps

The primary fermentation

1. Peel the skin off mangoes and remove the flesh, then add to the fermentor along with white grape concentrate, sugar, acid blends and hot water. Stir vigorously until the sugar dissolves completely.
2. Peel Add yeast nutrients, pectin enzyme, tannin, campden tablets and cold water to step 1. Stir well, then cover the fermentor with a lid and let sit for 1 hour.
3. Peel Rehydrate yeast with 1 cup of lukewarm water (38℃). Let sit for 10 to 30 minutes and add to fermentor. Stir well.
4. Peel Cover the fermentor, or seal loosely with plastic wrap. Let sit in a cool, dark corner at around 23℃.
5. Peel Let sit for 24 hours, then open to examine. If foam has formed on the surface, fermentation has started. If not, wait for another day and check again. If there is still no foam on the surface, refer to Question & Answer 1 on p.111 〔Wine Making Q&A〕.
6. Peel Stir twice a day, once in the morning and again at night, with a long stirring paddle.

W
winemaking at home

第二次發酵

7. 每隔1天，取一些酒醪，用比重測糖器檢測糖度，當含糖量降到5度以下時，用細網過濾出果汁，丟棄果肉渣，將酒醪換入第二次發酵用的果酒瓶內，加入3片二氧化硫片（先壓碎，以少許冷水溶解）。用橡皮塞封口加上發酵鎖，放在20°C以下的陰暗處。
8. 靜置10天後虹吸換瓶，果酒瓶要裝滿避免有空氣，引起氧化。
9. 靜置3週後，再次虹吸換瓶，加入皂土澄清。
10. 10天後虹吸換瓶，加入3片二氧化硫片（先壓碎，以少許冷水溶解）。
11. 靜置1個月。
12. 虹吸換瓶，或用細網過濾，加入200c.c.安定劑。
13. 裝瓶後，繼續熟成5個月。

The secondary fermentation

7. Remove a little must from the fermentor every other day and measure the gravity of the sugar with a hydrometer. When the sugar gravity is below 5, use a fine strainer to remove the particles, then transfer the must to the bottles for the secondary fermentation. Add 3 campden tablets (crush first and dissolve in a little cold water). Seal with rubber bungs and fit with airlocks. Transfer to a cool, dark corner with the temperature below 20°C.
8. Let sit for 10 days, then siphon and rack to bottles. Be sure to fill the bottles completely, so that there is no air, which might cause oxidization.
9. Let sit for 3 weeks, then siphon and rack to bottles along with bentonite to clear the must.
10. Let sit for 10 days, then siphon and rack to bottles once again. This time add 3 campden tables (crushed first and dissolved in a little cold water).
11. Let sit for 1 month.
12. Siphon and rack to bottles again, or strain off with fine strainer, then add 200c.c. stabilizer.
13. Bottle. Continue aging for 5 months.

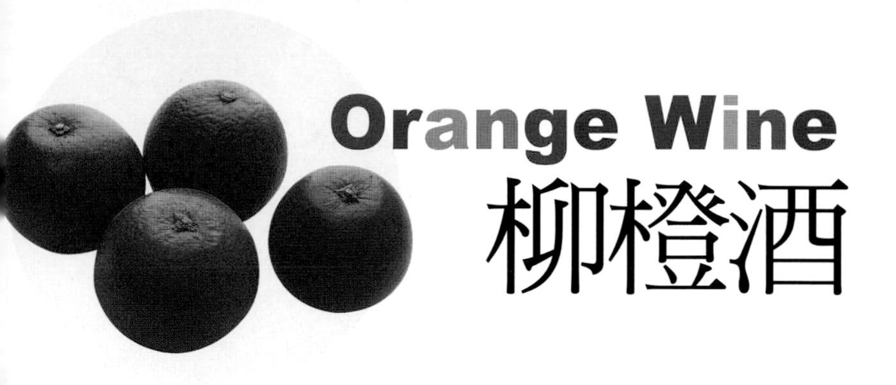

Orange Wine 柳橙酒

■ 製成份量：約20公升
■ 完成時間：6個月可裝瓶

美國主婦們傳統的秘方，是用柳橙釀製高酒精度的柳橙酒再加入苦艾酒草藥，自製美味的開胃酒。台灣不容易買到苦艾酒草藥，所以加入薄荷來提味。

The traditional secret used by American housewives is to use oranges and vermouth herbs to make strong orange wine and serve as an appetizer wine. In Taiwan, it is difficult to find vermouth herbs, therefore mint is added to enhance the flavor.

材料1：（第一次發酵）

柳橙汁	20公升
柳橙	4粒
葡萄乾	1公斤
酵母營養劑	2小匙
果膠酵素	2小匙
單寧	3小匙
二氧化硫片（壓碎）	5片
水果酒酵母	1包（5公克）
溫水	1杯

材料2：（過程中添加）

果糖	900 c.c.
皂土	2公克
二氧化硫片	3片
安定劑	300c.c.
新鮮薄荷	約38公克

Ingredients 1（The primary fermentation）
20 liters orange juice
4 oranges
1 kgs. raisin
2 tsp. yeast nutrients
2 tsp. pectin enzyme
3 tsp. tannin
5 campden tablets(crushed)
1 pack fruit wine yeast(5 grams)
1 cup lukewarm water

Ingredients 2
（Add during the process）
900grams sugar
2 grams bentonite
3 campden tablets
300c.c. stabilizer
1.3 oz. mint

■ Finished Amount : about 20 liters
■ Finishing Time : 6 months before bottling

柳橙

柳橙汁

單寧粉

二氧化硫

葡萄乾

糖

水果酒酵母

果膠酵素

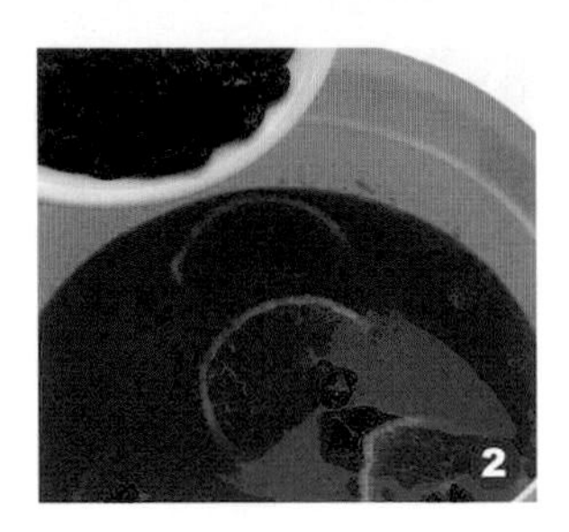

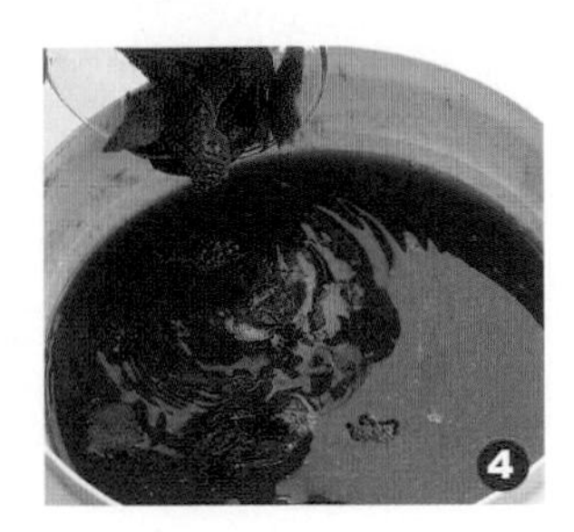

做法

第一次發酵

1. 用比重測糖器檢測柳橙汁糖度，參考P.103技術篇增糖之方法，將糖度提高到24度。
2. 4粒柳橙連皮切薄片，葡萄乾切碎，與糖度24度的柳橙汁一起放入第一次發酵桶。
3. 加入酵母營養劑、果膠酵素、單寧及二氧化硫，蓋好發酵桶，靜置1小時。
4. 酵母加入1杯溫水中（38°C），放置10～30分鐘，倒入發酵桶，用力攪拌。
5. 蓋好發酵桶，放置在23°C左右的陰暗處。
6. 待24小時後檢查，見泡沫浮在表面，表示發酵作用已經開始；若沒有發酵現象，再等24小時，若仍沒有動靜，請參考P.111（釀酒Q&A）問題1的原因解答。
7. 每天用長柄匙，早晚各攪拌一次。

Steps

The primary fermentation

1. Measure the sugar gravity of the orange juice with a hydrometer. See instructions in theTechniques Chapter on p.102. Increase the sugar gravity to 24.
2. Cut the four oranges into thin slices with the skin on. Crush raisins finely. Transfer both to the fermentor along with the orange juice, in which the gravity has already been increased.
3. Add yeast nutrients, pectin enzyme, tannins and campden tablets. Cover with a lid and let sit for 1 hour.
4. Re-hydrate yeast with 1 cup of lukewarm water (38°C). Let sit for 10 ~ 30 minutes and pitch into the fermentor. Stir vigorously until dissolved.
5. Replace the lid on the fermentor and remove it to a cool dark corner where the temperature is always around 23°C.
6. Let sit for 24 hours, then open to examine. If foam has formed on the surface, fermentation has started. If not, wait for another day and check again. If there is still no foam on the surface, refer to Question & Answer 1 on p.111〔Wine Making Q&A〕.
7. Stir twice a day, once in the morning and again at night, with a long stirring paddle.

❶ 柳橙切片加入柳橙汁中

❷ 柳橙酒中加入葡萄乾

❸ 虹吸時，利用高度的差異才能順利進行

❹ 柳橙酒中加入香草

W

winemaking at home

第二次發酵

8. 每隔1天，取一些酒醪，用比重測糖器檢測糖度，當含糖量降到5度以下時，用細網過濾出果汁，丟棄皮渣，將酒醪換入第二次發酵用的果酒瓶，用橡皮塞封口，加發酵鎖，放置在18°C左右的陰暗處。
9. 10天後虹吸換瓶，加入450c.c.果糖。
10. 10天後再加入450c.c.果糖，搖晃果酒瓶幫助果糖和酒醪混合。
11. 3週後虹吸換瓶，加入皂土澄清。
12. 10天後虹吸換瓶，加入薄荷，靜置1個月。
13. 虹吸後過濾，加入3片二氧化硫片（先壓碎，以少許冷水溶解）。
14. 靜置在18°C左右的陰暗處3個月。
15. 加入安定劑，裝瓶、繼續熟成6個月。

The secondary fermentation

8. Remove a little must from the fermentor every other day and measure the gravity of the sugar with a hydrometer. When the sugar gravity is below 5, use a fine strainer to remove the solids, then transfer the must to the bottles for the secondary fermentation. Seal with rubber bungs and fit with airlocks. Remove to a cool dark corner where the temperature does not exceed 18°C.
9. Let sit for 10 days, siphon and rack to bottles, add 450 grams sugar.
10. Let sit for 10 more days, add 450grams sugar fructose again, shake the bottles to help the sugar dissolve thoroughly into the must.
11. Wait for 3 weeks and siphon, then rack to bottles. Add bentonite to clear the must.
12. Let sit for 10 days, siphon and rack to bottles, add mint and wait for 1 month.
13. Siphon, strain and rack to bottles. Add 3 campden tablets (crushed first, then dissolved in a little cold water).
14. Let sit in a cool, dark corner, with the temperature maintained around 18°C for 3 months.
15. Add stabilizer, and bottle. Age for 6 months.

Pineapple Wine 鳳梨酒

■ 製成份量：約20公升
■ 完成時間：5個月可裝瓶

材料1：（第一次發酵）

鳳梨	4公斤
白葡萄濃縮汁	1公升
（可以新鮮青葡萄去皮榨汁取代）	
糖	4.5公斤
增酸劑	4小匙
熱開水	6公升
酵母營養劑	2小匙
果膠酵素	2小匙
單寧	2小匙
二氧化硫片（壓碎）	6片
冷水	8公升
水果酒酵母	1包（5公克）
溫水	1杯

材料2：（過程中添加）

果糖	500c.c.
皂土	2公克
二氧化硫片	3片
安定劑	300c.c.

剛釀好的鳳梨酒有很濃的鳳梨果香，經過熟成果香變淡，有點像白葡萄甜酒。

Just bottled pineapple wine has a thick pineapple aroma. The fruit aroma will become lighter after aging, similar to white grape wines.

■ Finished Amount : about 20 liters
■ Finishing Time : 5 months before bottling

Ingredients 1 (The primary fermentation)

4 kgs. pineapple
1 kgs. green grapes juice concentrate
4.5 kgs sugar
4 tsp. acid blend
6 liters hot water
2 tsp. yeast nutrients
2 tsp. pectin enzyme
2 tsp. tannin
6 tablets sulfur dioxide(crushed)
8 liters cold water
1 pack fruit wine yeast(5 grams)
1 cup lukewarm water

Ingredients 2 (Add during the process)

500grams sugar
2 grams bentonite
3 campden tablets
300c.c. stabilizer

鳳梨

白葡萄濃縮汁

二氧化硫

水果酒酵母

糖

單寧粉

果膠酵素

做法

第一次發酵

1. 鳳梨去皮、切碎，白葡萄濃縮汁、增酸劑、糖、熱開水放入第一次發酵桶。攪拌到糖全部溶化，也可以先用開水溶糖，再倒入發酵桶。
2. 加入酵母營養劑、果膠酵素、單寧、二氧化硫、及冷水，蓋好發酵桶，放置1小時。
3. 酵母加入1杯溫水中（38°C）放置10～30分鐘，倒入發酵桶中拌勻。
4. 蓋好發酵桶，放置在23°C左右的陰暗處。
5. 待24小時後檢查，見果肉浮在表面，表示發酵作用已經開始；若沒有發酵現象，再等一天，若仍沒有動靜，請參考P.111（釀酒Q&A）問題1的原因解答。
6. 每天用長柄匙，早晚各攪拌一次。

Steps

The primary fermentation

1. Peel the skin off from the pineapples and crush the flesh finely. Pour into the fermentor along with grape juice concentrate, acid blends, sugar and hot water for the first fermentation. Stir vigorously until the sugar dissolves completely, or let it dissolve in hot water first before pouring into the fermentor.
2. Add yeast nutrients, pectin enzyme, tannin, campden tablets and cold water. Cover with the lid and sit for 1 hour.
3. Rehydrate yeast in 1 cup of lukewarm water (38°C), sit for 10 to 30 minutes, then pour into the fermentor and mix well.
4. Cover the fermentor and transfer to a dark, cool corner with the temperature maintained at around 23°C.
5. Let sit for 24 hours, then open to examine. If foam has formed on the surface, fermentation has started. If not, wait for another day and check again. If there is still no foam on the surface, refer to Question & Answer 1 on p.111〔Wine Making Q&A〕.
6. Stir twice a day, once in the morning and again at night, with a long stirring paddle.

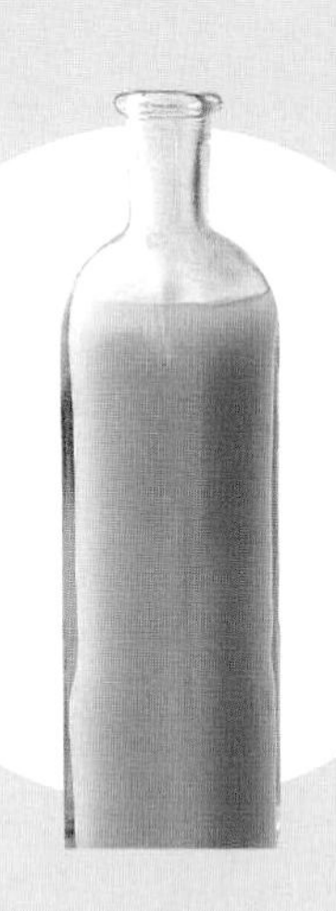

❶ 鳳梨切塊
❷ 切塊鳳梨加水及其他原料
❸ 用力吸一口，將果汁或酒醪引出
❹ 比重測糖器在測果汁時顯出之糖度

第二次發酵

7. 每隔1天，取出一些酒醪，用比重測糖器檢測糖度，當含糖量降到5度以下後，用細網過濾出鳳梨果汁，丟棄果肉渣。
8. 酒醪裝入第二次發酵用的果酒瓶，用橡皮塞封口，加發酵鎖，放置在陰暗處。
9. 10天後虹吸換瓶，加入500c.c.果糖，用橡皮塞封口，加發酵鎖，放置在陰暗處。
10. 靜置1個月，虹吸換瓶，加入皂土澄清。
11. 10天後虹吸換瓶，用細網過濾。
12. 加入3片二氧化硫片（先壓碎，以少許冷水溶解）。
13. 靜置3個月。
14. 虹吸換瓶，加入300 c.c.安定劑。
15. 裝瓶後，繼續熟成6個月。

The secondary fermentation

7. Remove a little must from the fermentor to measure the sugar gravity. If the sugar gravity is below 5, strain with a fine strainer to remove the solid particles.
8. Rack the must to bottles for the second fermentation. Seal with rubber bungs and fit with airlocks. Transfer the bottles to cool, dark corner.
9. Let sit for 10 days, siphon and rack to bottles, then add 500grams sugar. Seal the opening with rubber bungs and fit with airlocks. Let sit in dark, cool corner.
10. Let sit for 1 month, siphon and rack to bottles, then add bentonite to clear the must.
11. Let sit for 10 days, siphon and rack to bottles, then strain with a fine strainer to remove any solids.
12. Add 3 campden dioxide tablets (crushed first, then dissolved in a little cold water).
13. Let sit for 3 months.
14. Siphon and rack to bottles, add 300c.c. stabilizer to mix.
15. Bottle and let age for 6 months.

Pitahaya Wine
火龍果酒

■ 製成份量：約20公升
■ 完成時間：5個月可裝瓶

火龍果的果肉有白色和紅色兩種，都是釀酒的好原料。但火龍果的含酸度很低，製酒時要把酸度調高，才能釀出清淡可口、獨具特色的美酒。

The flesh of pitahaya comes in two different colors, white and red. Both are good for making wine. However, pitahaya is low in acid, so the acid gravity has to be increased to give the wine its signature light and delicious flavor.

材料1：（第一次發酵）

去皮切塊紅龍果肉	7公斤
糖	4.5公斤
增酸劑	15小匙
熱開水	6公升
酵母營養劑	2小匙
果膠酵素	2小匙
單寧	2小匙
二氧化硫片（壓碎）	6片
冷水	8公升
水果酒酵母	1包（5公克）
溫水	1杯

材料2：（過程中添加）

吉利丁粉	5公克
二氧化硫片	3片
安定劑	300c.c.

Ingredients 1 (The primary fermentation)

7 kgs. peeled pitahaya
4.5 kgs. sugar
15 tsp. acid blend
6 liters hot water
2 tsp. yeast nutrients
2 tsp. pectin enzyme
6 campden tablets(crushed)
8 liters cold water
1 pack fruit wine yeast(5 grams)
1 cup lukewarm water

Ingredients 2 (Add during the process)

5 grams gelatin powder
3 campden tablets
300c.c. stabilizer

■ Finished Amount : about 20 liters
■ Finishing Time : 5 months before bottling

火龍果

水果酒酵母

二氧化硫

糖

單寧粉

果膠酵素

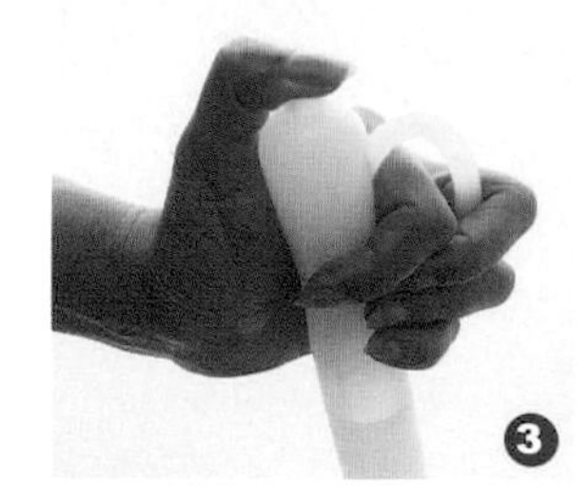

❶ 火龍果切塊
❷ 添加酵母營養劑
❸ 使用取酒器，用拇指將上端開口封住

做法

第一次發酵

1. 紅龍果肉切塊、壓碎，將糖、熱開水、增酸劑放入第一次發酵桶，攪拌到糖全部溶化。
2. 加入酵母營養劑、果膠酵素、單寧、二氧化硫粉末及冷水，攪拌均勻，蓋好發酵桶，放置1小時。
3. 酵母加入1杯溫水中（38℃）放置10～30分鐘，倒入發酵桶中拌勻。
4. 蓋好發酵桶，放置在23℃左右的陰暗處。
5. 待24小時後檢查，見果肉及泡沫浮在表面，表示發酵作用已經開始；若沒有發酵現象，再等一天，若仍沒有動靜，請參考P.111（釀酒Q&A）問題1的原因解答。
6. 每天用長柄匙，早晚各攪拌一次。

Steps

The primary fermentation

1. Cut peeled pitahaya into small pieces, crush and add to the fermentor along with sugar, hot water and acid blend for the primary fermentation. Stir vigorously until the sugar dissolves completely.
2. A dd yeast nutrients, pectin enzyme, tannin, crushed campden tablets and cold water to method (1). Stir until evenly mixed. Cover with the lid and let sit for 1 hour.
3. Rehydrate yeast with 1 cup of lukewarm water (38℃). Sit for 10 to 30 minutes, then pitch into the fermentor.
4. Cover the fermentor with a lid and put in a dark, cool location with the temperature maintained at around 23℃.
5. Let sit for 24 hours, then open to examine. If foam has formed on the surface, fermentation has started. If not, wait for another day and check again. If there is still no foam on the surface, refer to Question & Answer 1 on p.111〔Wine Making Q&A〕.
6. Stir twice a day, once in the morning and again at night, with a long stirring paddle.

◎釀酒TIPS
紅龍果酒容易氧化，所以在裝入第二次發酵用的果酒瓶時，就要先加入二氧化硫來預防氧化。

Winemaking Tips
Because pitahaya oxidizes easily, campden tablets are added during the first step of secondary fermentation.

winemaking at home

第二次發酵

7. 每隔1天，取出一些酒醪，用比重測糖器檢測糖度，當含糖量降到5度以下後，用細網濾出果汁，丟棄果肉渣。
8. 酒醪裝入第二次發酵用的果酒瓶，加入3片二氧化硫片（先壓碎，以少許冷水溶解），用橡皮塞封口，加發酵鎖，放置在約18℃左右陰暗處。
9. 10天後虹吸換瓶。
10. 靜置1個月，虹吸換瓶，加入吉利丁粉澄清。
11. 10天後虹吸換瓶，加入300 c.c.安定劑，靜置1個月。
12. 裝瓶後，繼續熟成2個月。

The secondary fermentation

7. Remove a little must from the fermentor and measure the sugar gravity with a hydrometer. If the sugar gravity is below 5, strain the must with a strainer to remove any solid particles.
8. Rack the must to bottles for the secondary fermentation along with 3 campden tablets (crushed first, then dissolved in a little water). Seal with rubber bungs and fit with airlocks. Transfer to a cool, dark location with the temperature maintained at around 23℃.
9. Let sit for 10 days, siphon and rack to bottles.
10. Let sit for 1 month, siphon and rack to bottles along with gelatin powder added to clear the must.
11. Let sit for 10 days, siphon and rack to bottles along with 300c.c. of stabilizer added. Let sit for 1 month.
12. Bottle and let age for 2 months.

Plum Wine
梅酒

■ 製成份量：約20公升
■ 完成時間：3個月可裝瓶

4月開始，市場上就可以看到黃梅的蹤跡了，米酒加冰糖所浸泡出的黃梅酒是台灣主婦的拿手私釀酒。這兒介紹一種不一樣的釀法，更清爽可口，好喝的讓你一杯接一杯。

Plums can be seen in markets after April. Plums soaked in rice wine and rock sugar is a specialty of Taiwanese housewives. Another method of making plum wine, which is lighter, delicious, and irresistible, is presented here.

材料1：（第一次發酵）

熟黃梅	10公斤
糖	6公斤
水	10公升
酵母營養劑	2小匙
果膠酵素	2小匙
單寧	2小匙
碳酸鈣（壓碎）	20公克
水果酒酵母	1包（5公克）
溫水	1杯

材料2：（過程中添加）

吉利丁粉	5公克
二氧化硫片	3片
安定劑	500c.c.

Ingredients 1（The primary fermentation）

10 kgs. ripe yellow plums
6 kgs. sugar
10 liters water
2 tsp. yeast nutrients
2 tsp. pectin enzyme
2 tsp. tannin
20 grams calcium carbonate(crushed)
1 pack fruit wine yeast(5 grams)
1 cup lukewarm water

Ingredients 2（Add during the process）

5 grams gelatin powder
3 campden tablets
500c.c. stabilizer

■ Finished Amount : about 20 liters
■ Finishing Time : 3 months before bottling

熟黃梅

水果酒酵母

二氧化硫

糖

單寧粉

果膠酵素

❶ 比重測糖器在酒醪中顯示之含糖度

❷ 酒醪在第二次發酵果酒瓶中加橡皮塞及發酵鎖

❸ 梅酒，澄清中

做法

第一次發酵

1. 將10公升水煮開，加入糖煮至融化，熄火；放入第一次發酵桶。
2. 黃梅挑除腐爛及不熟的青梅。
3. 將黃梅與酵母營養劑、果膠酵素、單寧、碳酸鈣粉末一起倒入發酵桶，攪拌均勻，蓋好發酵桶，放置1小時。
4. 酵母加入1杯溫水（38℃），放置10～30分鐘；當發酵桶內溫度降至與體溫相同，約38℃時，將酵母水倒入發酵桶內拌勻。
5. 蓋好發酵桶，放置在23℃左右的陰暗處。
6. 待24小時後檢查，見泡沫浮在表面，表示發酵作用已經開始；若沒有發酵現象，再等一天，若仍沒有動靜，用溫水加入1包酵母及2小匙酵母營養劑，倒入發酵桶。
7. 每天用長柄匙，早晚各攪拌一次。

Steps

The primary fermentation

1. Bring 10 liters of water to a boil in a large pot. Add sugar and cook until sugar dissolves completely, remove from heat. Transfer to a fermentor for the primary fermentation.
2. Discard any spoiled or any unripe plums.
3. Add plums along with yeast nutrients, pectin enzyme, tannin and calcium carbonate in the fermentor in step 1. Stir until evenly mixed. Cover with the lid and let sit for 1 hour.
4. Rehydrate the yeast with 1 cup of lukewarm water (38℃). Let sit for 10 to 30 minutes, wait until the temperature in the fermentor falls to 38℃, then pour dissolved yeast into the fermentor.
5. Cover the fermentor with a lid and move to a dark, cool location with the temperature maintained at around 23℃.
6. Let sit for 24 hours, then open to examine. If foam has formed on the surface, fermentation has started. If not, wait for another day and check again. If there is still no foam, re-hydrate another pack of yeast and 2 tsp. of yeast nutrients in lukewarm water, pour into the fermentor again.
7. Stir twice a day, once in the morning and again at night, with a long stirring paddle.

◎釀酒TIPS
梅子比較酸，所以要加入碳酸鈣來降酸。

Winemaking Tips
Plum is somewhat sour, therefore calcium carbonate is added to decrease the acidity.

W
winemaking at home

第二次發酵

8. 5天後，壓爛黃梅，將酒醪全部換入第二次發酵果酒瓶，用橡皮塞封口，加發酵鎖，放置在23°C左右的陰暗處。
9. 10天，虹吸換瓶，加入3片壓碎的二氧化硫片（先用少許冷水溶解），橡皮塞封口，加發酵鎖。
10. 1個月後，虹吸換瓶加入吉利丁粉澄清。
11. 10天後虹吸換瓶，過濾、加入500c.c.安定劑，靜置1個月。
12. 裝瓶即可飲用，但繼續熟成2個月後更順口。

The secondary fermentation

8. Let sit for 5 days, then crush and mash the plums. Strain the must with a strainer and rack to bottle for the second fermentation. Seal with rubber bungs and fit with airlocks. Move bottles to a cool, dark corner with the temperature maintained at around 23°C.
9. Let sit for 10 days, siphon and rack to bottles along with 3 crushed campden tablets added (dissolved in a little water). Seal with rubber bungs and fit with airlocks.
10. Let sit for 1 month, siphon and rack to bottles along with gelatin powder added to clear the must.
11. Let sit for 10 days, siphon and rack to bottles along with 500c.c. of stabilizer added. Let sit for 1 month.
12. Bottle and let age for 2 months for a better flavor.

四季新鮮水果酒

台灣一年四季產有各種不同新鮮水果，都是釀酒的上好原料。水果豐收時，沒有時間釀酒，只要將新鮮水果放到二氧化硫水溶液裡浸一下（250PPM），再用冷凍袋裝好放在冰凍櫃裡，等到有時間時再拿出來釀酒，冷凍過的水果因為細胞壁已破壞，釀酒時出汁率比新鮮水果多，第一次發酵時，也不必再添加二氧化硫。冷凍水果和新鮮水果釀出來的酒風味相同，沒有什麼差別。

Four-Season Fresh Fruit Wine

There are many different kinds of fresh fruit available in Taiwan and they are all good ingredients for winemaking . During the fruit season, if you have no time to make wine, no problem, soak the fresh fruit in sulfur dioxide solution (250PPM) for a second, then seal in freezer bag and freeze until ready to use. Remove the frozen fruit and start the winemaking process. Because cell walls are destroyed by freezing, the fruit should give up its water at a faster rate than its fresh counterpart. The campden tablets are not needed for the primary fermentation. Wine tastes the same whether it is made using fresh or frozen fruit.

Apple Wine
蘋果酒

■ 製成份量：約25公升
■ 完成時間：3個月可裝瓶

蘋果採收後，不要立刻拿來釀酒，讓蘋果放置幾天，繼續成熟，甜度會增加，壓汁時的出汁率比較多，下面這個方法釀出的蘋果酒，清淡香甜。

Do not use apples right after they are picked. Let the apples ripen for a couple of days. They will naturally continue ripening, so the sweetness will increase and they will release more liquid. Follow every step in this recipe and the apple wine will be clear and sweet.

材料1：（第一次發酵）

蘋果	14公斤
二氧化硫片	8 1/2片
冷水	3杯
糖	5公斤
增酸劑	2小匙
熱開水	6公升
酵母營養劑	2小匙
果膠酵素	4小匙
單寧	3小匙
冷水	10公升
香檳酒酵母	1包（5公克）
溫水	1杯

材料2：（過程中添加）

二氧化硫片	6片
吉利丁粉	5公克
安定劑	300c.c.

Ingredients 1（The primary fermentation）

14 kgs. apples
8 1/2 campden tablets(crushed)
3 cups cold water
5 kgs. sugar
2 tsp. acid blend
6 liters hot water
2 tsp. yeast nutrients
4 tsp. pectin enzyme
3 tsp. tannin
10 liters cold water
1 pack champagne yeast(5 grams)
1 cup lukewarm water

Ingredients 2（Add during the process）

6 campden tablets(crushed)
5 grams gelatin powder
300c.c. stabilizer

■ Finished Amount : about 25 liters
■ Finishing Time : 3 months before bottling

蘋果

香檳酒酵母

二氧化硫

糖

單寧粉

果膠酵素

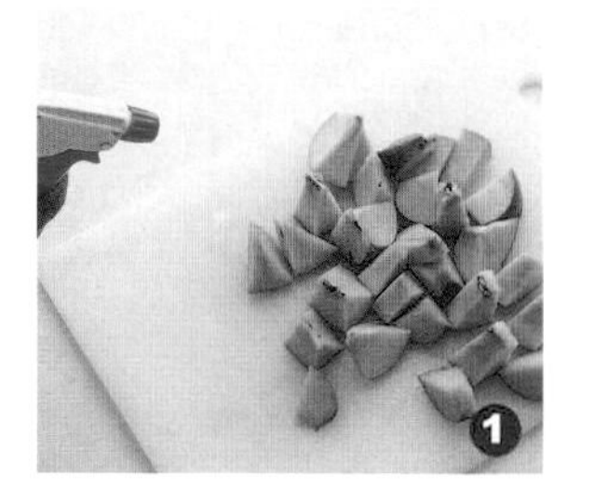

❶ 蘋果切塊，噴二氧化硫水，避免氧化變褐色

❷ 添加單寧

❸ 吉利丁加入冷水中溶解

做法

第一次發酵

1. 挑選熟透而沒有腐爛的蘋果。
2. 將8片二氧化硫片溶在3杯冷水裡，裝入噴水瓶。
3. 蘋果切小塊，切時要噴灑二氧化硫水，避免蘋果褐變。
4. 將蘋果塊、熱開水、糖、增酸劑倒入第一次發酵桶，攪拌到糖全部溶解。
5. 加入酵母營養劑、單寧、果膠酵素、冷水，攪拌混合，蓋好發酵桶，放置1小時。
6. 酵母加入1杯溫水（38°C），10～30分鐘後，加入發酵桶中，用力攪拌均勻，蓋好發酵桶。
7. 放置在23°C左右陰暗處，待24小時後檢查，見果肉及泡沫浮在表面，表示發酵作用已經開始；若沒有發酵現象，再等24小時，若仍沒有動靜，請參考P.111（釀酒Q&A）問題1的原因解答。
8. 每天用長柄匙，早晚各攪拌一次。

Steps

The primary fermentation

1. Select perfect ripe apples without any spoilage.
2. Dissolve 8 campden tablets with 3 cups of cold water, then transfer the solution to a spray bottle.
3. Cut apples into small pieces. Spray all cut apples with sulphite solution to prevent darkening.
4. Transfer apple pieces, hot water, sugar, acid blend to a fermentor. Stir until sugar dissolves thoroughly.
5. Add yeast nutrients, pectin enzyme, tannin, and cold water. Stir until evenly mixed. Cover with the lid and let sit for 1 hour.
6. Rehydrate yeast with 1 cup of lukewarm water (38°C). Sit for 10 to 30 minutes, then pour into the fermentor.
7. Cover the fermentor with a lid and move to a dark and cool location with the temperature maintained at around 23°C.
8. Let sit for 24 hours, then open to examine. If foam has formed on the surface, fermentation has started. If not, wait for another day and check again. If there is still no foam on the surface, refer to Question & Answer 1 on p.111〔Wine Making Q&A〕.

W

winemaking at home

第二次發酵

9. 每隔1天，取出一些酒醪，用比重測糖器檢測糖度，當含糖量降到5度以下時，不再攪動酒醪，將上層汁液虹吸到第二次發酵用的果酒瓶，丟棄底層的蘋果及沈澱；果酒瓶用橡皮塞封口，加上發酵鎖，放置在18℃的陰暗處。

10. 10天後虹吸換瓶，加入3片二氧化硫（先壓碎，以少許冷水溶解）。

11. 靜置1個月，虹吸換瓶，加入吉利丁粉澄清。

12. 10天後虹吸，過濾。蘋果酒容易氧化，此時再加入3片二氧化硫。

13. 靜置1個月，虹吸換瓶，加入安定劑300c.c.。

14. 裝瓶後，繼續熟成4個月。

The secondary fermentation

9. Stir twice a day, once in the morning and again at night, with a long stirring paddle.

10. Remove a little must from fermentor and measure the sugar gravity with a hydrometer every other day. If the sugar gravity is below 5, do not stir the must anymore. Siphon the upper layer of the must to bottles for the secondary fermentation. Discard the bottom solids and sediment. Seal with rubber bungs and fit with airlocks. Move to a cool, dark location with the temperature maintained at around 18℃.

11. Let sit for 10 days, siphon and rack to bottles along with 3 sulfur dioxide tablets added (crushed first, then dissolved in a little cold water).

12. Let sit for 1 month, siphon and rack to bottles along with gelatin powder added to clear the must.

13. Let sit for 10 days, siphon and rack to bottles. Apple wine oxidizes very easily, so add 3 campden tablets to mix.

14. Let sit for 1 month, siphon and rack to bottles, along with 300c.c. of stabilizer added.

15. Bottle and age for 4 months.

Apple Cider Wine
蘋果汽泡酒

■ 製成份量：約25公升
■ 完成時間：1個月可裝瓶

材料1：（第一次發酵）

蘋果	60公斤
果膠酵素	2小匙
酵母營養劑	2小匙
單寧	2小匙
水果酒酵母	1包（5公克）
溫水	1杯

材料2：（過程中添加）

吉利丁粉	5公克
二氧化硫片	3片
糖	200公克
香檳酒酵母	1包（5公克）

蘋果樹是歐美最普遍的果樹，就像台灣的番石榴樹，到處可見。烤蘋果派、釀蘋果酒是主婦們最拿手的，蘋果汽泡酒顏色混濁，口感粗獷，充滿了汽泡，頗有鄉村野趣。

Apple trees are the most common trees in Europe and America, just as guava trees are in Taiwan. Filled with countryside fun, apple cider wine has a cloudy color, and the texture is quite thick and full of soda bubbles.

Ingredients 1 (The primary fermentation)
60 kgs. apples
2 tsp. pectin enzyme
2 tsp. yeast nutrients
2 tsp. tannin
1 pack fruit wine yeast(5 grams)
1 cup lukewarm water

Ingredients 2 (Add during the process)
5 grams gelatin powder
3 tablets sulfur dioxide
200 grams sugar
1 pack champagne yeast(5 grams)

■ Finished Amount : about 25 liters
■ Finishing Time : 1 months before bottling

蘋果

水果酒酵母

二氧化硫

糖

單寧粉

果膠酵素

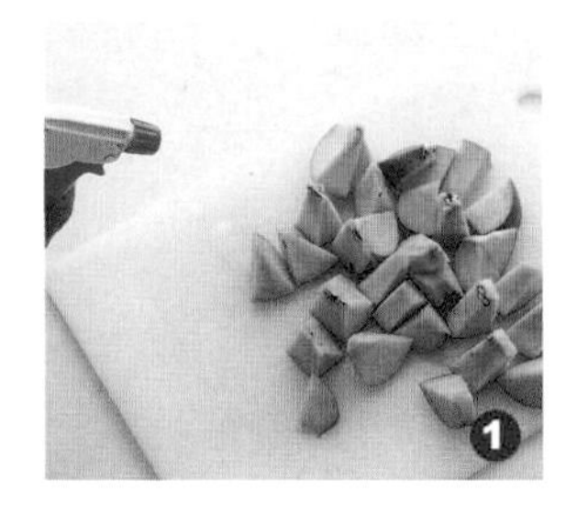

❶ 蘋果切塊，噴二氧化硫水，避免氧化變褐色

❷ 蓋緊發酵桶

❸ 用豆漿袋（紗網）擠出果汁

做法

第一次發酵

1. 挑選熟透而沒有腐爛的蘋果。
2. 將8片二氧化硫片溶在3杯冷水裡，裝入噴水瓶。
3. 蘋果切小塊，切時要噴灑二氧化硫水，避免蘋果褐變。
4. 將蘋果塊放入第一次發酵桶，加入酵母營養劑、單寧、果膠酵素、冷水，攪拌均勻，蓋好發酵桶，放置1小時。
5. 酵母加入1杯溫水（38°C），10～30分鐘後，加入發酵桶中，用力攪拌，蓋好發酵桶，放置在23°C左右陰暗處。
6. 待24小時後檢查，見汽泡自桶底昇到表面，表示發酵作用已經開始；若沒有發酵現象，再等24小時，若仍沒有動靜，請參考P.111（釀酒Q&A）問題1的原因解答。
7. 每天用長柄匙，早晚各攪拌一次。

Steps

The primary fermentation

1. Select ripe apples without any spoilage.
2. Dissolve 8 campden tablets with 3 cups of cold water, then transfer the solution to a spray bottle.
3. Cut apples into small pieces. Spray all cut apples with sulphite solution to prevent darkening.
4. Transfer the apple pieces to a fermentor along with yeast nutrients, tannin, pectin enzyme and cold water. Mix well until even, then cover with a lid and let sit for 1 hour.
5. Rehydrate yeast with 1 cup of lukewarm water (38°C) for 10 ~ 30 minutes. Pour to the fermentor and stir vigorously. Put the lid back and move the fermentor to a dark andcool location with the temperature maintained at around 23°C.
6. Let sit for 24 hours, then open to examine. If foam has formed on the surface, fermentation has started. If not, wait for another day and check again. If there is still no foam on the surface, refer to Question & Answer 1 on p.111〔Wine Making Q&A〕.
7. Stir twice a day, once in the morning and again at night, with a long stirring paddle.

第二次發酵

8. 5天後用細網擠出果汁，丟棄蘋果皮渣等，將酒醪換入第二次發酵用的果酒瓶，用橡皮塞封口，加上發酵鎖，放在18℃的陰暗處。

9. 每隔1天，取一些酒醪，用比重測糖器檢測糖度，當含糖量降到3度以下時，虹吸換瓶。

10. 靜置15天，再次虹吸換瓶，加入吉利丁粉及3片二氧化硫片（先壓碎，以少許冷水溶解）。

11. 10天後，虹吸換瓶取出1大碗酒醪溶解200公克蔗糖及香檳酵母，倒入果酒瓶，用橡皮塞封口加發酵鎖。

12. 12小時後，裝瓶，要用耐壓力的瓶子，例如啤酒瓶、香檳瓶或汽水瓶，瓶塞要蓋緊。

13. 裝瓶後，繼續熟成3個月。

The secondary fermentation

8. Sit for 5 days, then strain the must with a strainer and discard all the solids. Rack the must to bottles for the secondary fermentation. Seal with rubber bungs and fit with airlock. Remove to cool corner with the temperature maintained at around 18°C.

9. a little must from the bottles every other day. Use a hydrometer to measure the sugar gravity. When it falls below 3, siphon and rack to new bottles. Seal with rubber bungs and fit with airlock.

10. Let sit for 15 days, siphon and rack to new bottles along with gelatin powder as well as 3 campden tablets (crushed first, then dissolved in a little cold water).

11. Let sit for 10 days, siphon 1 bowl of must and dissolve 200 grams of sugar and champagne yeast inside. Mix well and rack to bottles, seal with rubber bungs and fit with airlocks.

12. Sit for 12 hours, and bottle in pressure resistant bottle, such as beer bottle, champagne bottle or soda bottle. Seal tightly.

13. Let age for at least 3 months.

Guava Wine
芭樂酒

■ 製成份量：約25公升
■ 完成時間：5個月可裝瓶

泰國品種的芭樂及台灣本地的紅心芭樂，用同樣的方法釀酒，紅心芭樂酒比較香醇，泰國芭樂有一絲青澀味，將泰國芭樂削皮再切碎，減少了青澀味，效果不錯。珍珠芭樂則不必削皮，可直接切碎釀酒。

Guava from Thailand and red-flesh guava from local trees can be used the same way. Red-flesh guava wine is thicker and more fragrant. Thai guava wine has a dry taste. The second time I tried making wine from Thai guava, I peeled the skin first before cutting into pieces. The wine is mild and the overall effect is better. Pearl guava does not need to have the skin remove, just crush and use directly.

材料1：（第一次發酵）

芭樂	16公斤
糖	7公斤
白葡萄濃縮汁	1公升
(可以新鮮青葡萄去皮榨汁取代)	
增酸劑	8小匙
熱開水	6公升
酵母營養劑	2小匙
果膠酵素	2小匙
冷水	8公升
二氧化硫片	6片
香檳酒酵母	1包（5公克）
溫水	1杯

材料2：（過程中添加）

皂土	2公克
二氧化硫片	3片
安定劑	300c.c.

Ingredients 1 (The primary fermentation)

16 kgs. guava
7 kgs. sugar
1 kgs. green grapes juice concentrate
8 tsp. acid blend
6 liters hot water
2 tsp. yeast nutrients
2 tsp. pectin enzyme
8 liters cold water
6 campden tablets
1 pack champagne yeast(5 grams)
1 cup lukewarm water

Ingredients 2 (Add during the process)

2 grams bentonite
3 campden tablets
300c.c. stabilizer

■ Finished Amount : about 25 liters
■ Finishing Time : 5 months before bottling

芭樂

白葡萄濃縮汁

香檳酒酵母

二氧化硫

糖

果膠酵素

❶ 番石榴切塊
❷ 添加增酸劑
❸ 加入二氧化硫水溶液

做法

第一次發酵

1. 挑選熟透而沒有腐爛的芭樂，切碎；倒入第一次發酵桶，加入白葡萄濃縮汁、糖、熱開水、增酸劑，用力攪拌，直到糖溶解。
2. 將酵母營養劑、果膠酵素、冷水和二氧化硫依序加入發酵桶，攪拌混合，蓋好，放置1小時。
3. 酵母加入1杯溫水（38℃），10～30分鐘後，倒入發酵桶，攪拌混合，蓋好，放在23℃陰暗處。
4. 待24小時後檢查，見果肉及泡沫浮在表面，表示發酵作用已經開始；若沒有發酵現象，再等24小時；仍沒有動靜，請參考P.111（釀酒Q&A）問題1的原因解答。
5. 每天用長柄匙，早晚各攪拌一次。
6. 每隔1天，取一些酒醪，用比重測糖器檢測糖度。

Steps

The primary fermentation

1. Select extremely ripe guava without any spoilage and crush finely. Add to the fermentor along with white grape juice concentrate, sugar, hot water and acid blend added. Stir vigorously until sugar dissolves completely.
2. Add yeast nutrients, pectin enzyme, cold water and campden tablets in listed order to the fermentor. Stir frequently until mixed. Cover with the lid and let sit for 1 hour.
3. Rehydrate yeast with 1 cup of lukewarm water (38℃), sit for 10 to 30 minutes and pour into the fermentor. Stir well and put the lid back. Remove to a dark cool corner with the temperature maintained at around 23℃.
4. Let sit for 24 hours, then open to examine. If foam has formed on the surface, fermentation has started. If not, wait for another day and check again. If there is still no foam on the surface, refer to Question & Answer 1 on p.111 〔Wine Making Q&A〕.
5. Stir twice a day, once in the morning and again at night, with a long stirring paddle.
6. Remove a little must from the fermentor every other day and measure the sugar gravity with a hydrometer.

W

winemaking at home

第二次發酵

7. 當糖含量降到5度以下時，以細網擠出芭樂的果汁。
8. 將酒醪換入第二次發酵用的果酒瓶，用橡皮塞封口，加發酵鎖，放在18°C的陰暗處。
9. 10天後虹吸換瓶。
10. 1個月後虹吸換瓶，加入皂土澄清。
11. 10天後，虹吸過濾，加入3片二氧化硫片（先壓碎，以少許冷水溶解）。
12. 靜置3個月。
13. 虹吸換瓶加入300c.c.安定劑。
14. 裝瓶，繼續熟成6個月。

The secondary fermentation

7. When the sugar gravity is below 5, strain out the must with a fine strainer.
8. Rack the must to new bottles for the secondary fermentation. Seal with rubber bungs and fit with air-locks. Move the bottles to a cool and dark location with the temperature maintained at around 18 °C.
9. Sit for 10 days, siphon and rack to new bottles.
10. Sit for 1 month, siphon and rack to new bottles along with bentonite added.
11. Sit for 10 days more, siphon and rack to new bottles along with 3 campden tablets added (crushed first, then dissolved in a little cold water).
12. Let sit for 3 months.
13. Siphon and rack to new bottles along with 300c.c. stabilizer added.
14. Bottle and age for at least 6 months.

Strawberry Wine

草莓酒

■ 製成份量：約25公升
■ 完成時間：3個月可裝瓶

羅曼蒂克的玫瑰紅色，甜中有酸的清新口味，加點黑松汽水，冒泡的心情，好幸福喔！

A romantic rosy color, a sweet clear light flavor tinged with a little sour. Add a little soda and create a mood as exciting as the bubbles, how lucky we are!

材料1：（第一次發酵）

草莓	8公斤
紅葡萄酒濃縮汁	1公升
糖	4.5公斤
熱開水	6公升
增酸劑	4小匙
酵母營養劑	2小匙
單寧	2小匙
果膠酵素	2小匙
二氧化硫片	8片
冷水	8公升
水果酒酵母	1包（5公克）
溫水	1杯

材料2：（過程中添加）

皂土	2公克
二氧化硫片	3片
安定劑	300c.c.

Ingredients 1（The primary fermentation）

8kgs. strawberries
1 liters red grape wine concentrate
4.5kgs sugar
6 liters hot water
4 tsp. acid blend
2 tsp. yeast nutrients
2 tsp. tannin
2 tsp. pectin enzyme
8 campden tablets
8 liters cold water
1 pack fruit wine yeast(5 grams)
1 cup lukewarm water

Ingredients 2（Add during the process）

2 grams bentonite
3 campden tablets
300c.c. stabilizer

■ Finished Amount : about 25 liters
■ Finishing Time : 3 months before bottling

草莓

紅葡萄濃縮汁

二氧化硫

水果酒酵母

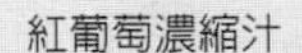

糖

單寧粉

果膠酵素

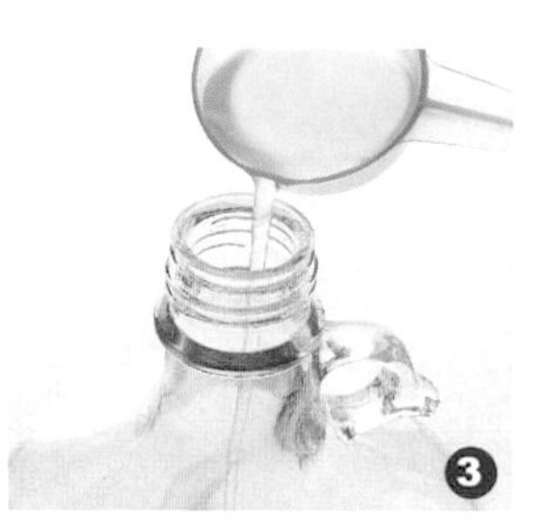

❶ 草莓去蒂
❷ 草莓壓爛
❸ 添加吉利丁

做法

第一次發酵

1. 選取紅熟的草莓，壓爛，倒入第一次發酵桶，加入糖，紅葡萄濃縮汁、熱開水、增酸劑攪拌到糖完全溶解。
2. 加入酵母營養劑、果膠酵素、單寧、冷水、二氧化硫片，攪拌均勻，蓋好，放置1小時。
3. 酵母加入1杯溫水（38℃），10～30分鐘後，倒入發酵桶攪拌均勻，放在23℃左右陰暗處。
4. 待24小時後檢查，見果肉及泡沫浮在表面，表示發酵作用已經開始；若沒有發酵現象，再等24小時；仍沒有動靜，請參考P.111（釀酒Q&A）問題1的原因解答。
5. 每天用長柄匙，早晚各攪拌一次。

Steps

The primary fermentation

1. Select red ripe strawberries, crush finely and transfer to the fermentor for the primary fermentation. Add sugar, red wine concentrate, hot water and acid, then stir until the sugar dissolves completely.
2. Continue to add yeast nutrients, pectin enzyme, tannin, cold water and cmapden tablets. Stir until well-mixed. Cover with a lid and sit for 1 hour.
3. Rehydrate yeast with 1 cup of lukewarm water (38℃) for 10 to 30 minutes. Pour into the fermentor and mix well. Put the lid back and move the fermentor to a cool and dark location with the temperature maintained at 23℃.
4. Let sit for 24 hours, then open to examine. If foam has formed on the surface, fermentation has started. If not, wait for another day and check again. If there is still no foam on the surface, refer to Question & Answer 1 on p.111〔Wine Making Q&A〕.
5. Stir twice a day, once in the morning and again at night, with a long stirring paddle.

W

winemaking at home

第二次發酵

6. 每隔1天，取一些酒醪，用比重測糖器檢測糖度。當糖含量降到5度以下時，以細網擠出果汁，小心地擠不要將細小的種子擠入酒醪。
7. 將酒醪換入第二次發酵用的果酒瓶，用橡皮塞封口，加發酵鎖，放在20℃陰暗處。
8. 10天後虹吸換瓶
9. 1個月後，虹吸換瓶，加入皂土澄清。
10. 10天後虹吸換瓶，加入二氧化硫3片（先壓碎，以少許冷水溶解）。
11. 1個月後，虹吸換瓶，加入300c.c.安定劑。
12. 裝瓶後，繼續熟成5個月。

The secondary fermentation

6. Remove a little must every other day and measure the sugar gravity with a hydrometer. If the sugar gravity is below 5, strain out the must with a fine strainer carefully, so as not to squeeze little seeds through into the must.
7. Rack to new bottles for the secondary fermentation. Seal with rubber bungs and fit with airlocks. Move bottles to cool and dark location with the temperature maintained at around 20℃.
8. Let sit for 10 days, siphon and rack to new bottles.
9. Let sit for 1 month, siphon and rack to new bottles. Add bentonite to clear the must.
10. Let sit for 10 days, siphon and rack to new bottles along with 3 campden tablets (crushed first, then dissolved with a little cold water).
11. Let sit for 1 month, siphon and rack to new bottles along with 300c.c. stabilizer added.
12. Bottle and continue aging for 5 months.

Grape Wine 紅葡萄酒

■ 製成份量：約15公升
■ 完成時間：8個月可裝瓶

台灣盛產葡萄，當水果食用的品種以巨峰、蜜紅這兩種為主，民間釀酒則以黑后為紅葡萄酒、金香葡萄為白葡萄酒的主要原料。本道酒譜採用市場上容易買到的巨峰或蜜紅葡萄來釀製紅葡萄酒。

Grapes are produced in large amounts in Taiwan. Two kinds of grape varieties served as fruits here are Chufeng and Mihung. Black queen is used in making red wine, while Chinhsiang grapes are the main ingredient in white wine. In this wine recipe, we utilize Chufeng and Mihung, which are easier to find in the market, to make red wine.

材料1：（第一次發酵）

紅葡萄	20公斤
糖	2.5公斤
酵母營養劑	2小匙
果膠酵素	2小匙
冷水	6公升
二氧化硫片	5片
紅葡萄酒酵母	1包（5公克）
溫水	1杯

材料2：（過程中添加）

吉利丁粉	5公克
二氧化硫片	6片
安定劑	300c.c.

Ingredients 1（The primary fermentation）

20 liters red grape
2.5kgs. sugar
2 tsp. yeast nutrients
2 tsp. pectin enzyme
6 liters cold water
5 campden tablets
1 pack red wine yeast(5 grams)
1 cup lukewarm water

Ingredients 2（Add during the process）

5 grams gelatin powder
6 campden tablets
300c.c. stabilizer

■ Finished Amount : about 15 liters
■ Finishing Time : 8 months before bottling

紅葡萄

紅葡萄酒酵母

二氧化硫

糖

吉利丁

果膠酵素

❶ 用手將葡萄擠破
❷ 發酵中的葡萄酒醪，葡萄皮肉均浮在表面
❸ 虹吸換瓶

做法

第一次發酵

1. 挑除壓爛及未熟的果粒。
2. 戴上清潔殺菌手套，用手將葡萄擠破，放入第一次發酵桶。
3. 加入酵母營養劑、果膠酵素、冷水、二氧化硫攪拌均勻，放置1小時。
4. 酵母加入1杯溫水（38℃），10～30分鐘後，倒入發酵桶攪拌均勻，蓋好，放在23℃左右陰暗處。
5. 待24小時後檢查，見果肉及泡沫浮在表面，表示發酵作用已經開始；若沒有發酵現象，再等24小時；仍沒有動靜，請參考P.111（釀酒Q&A）問題1的原因解答。
6. 每天早晚各一次，用長柄匙攪拌酒醪，把浮在表面的果肉及果皮壓到果汁中。
7. 每日取一些酒醪檢測糖度，3天後早晚各檢測一次，當含糖量降到5度以下時，立刻用細網過濾。

Steps

The primary fermentation

1. Discard any spoiled or unripened grapes.
2. Put on sanitary gloves and crush each grape by squeezing, then transfer to the fermentor for the primary fermentation.
3. Add yeast nutrients, pectin enzyme, and campden tablets to the fermentor. Let sit for 1 hour.
4. Rehydrate yeast with 1 cup of lukewarm water (38℃) for 10 to 30 minutes. Pour into the fermentor and stir until evenly mixed. Put the cover back and move to a cool and dark location with the temperature maintained at around 23℃.
5. Let sit for 24 hours, then open to examine. If foam has formed on the surface, fermentation has started. If not, wait for another day and check again. If there is still no foam on the surface, refer to Question & Answer 1 on p.111〔Wine Making Q&A〕.
6. Stir twice a day, once in the morning and again at night, with a long stirring paddle.
7. Remove a little must every other day to measure the sugar gravity with a hydrometer. When the sugar gravity is below 5, strain out the must with a fine strainer and remove the solids.

W

winemaking at home

第二次發酵

8. 擠出果肉中的汁液、果渣，丟棄不要。
9. 將酒醪換入第二次發酵用的果酒瓶，加入二氧化硫3片（先壓碎，以少許冷水溶解）。用橡皮塞封口，加發酵鎖，放在溫度約23°C的陰暗處。
10. 1個月後，虹吸換瓶，加入吉利丁澄清。
11. 10天後，再次虹吸換瓶，再次加入二氧化硫3片（先壓碎，以少許冷水溶解）。
12. 靜置6個月，放在溫度約23°C的陰暗處。
13. 虹吸過濾換瓶，加入安定劑即可裝瓶。
14. 裝瓶後，繼續熟成2個月。

The secondary fermentation

8. Rack the must to new bottles for the secondary fermentation. Add 3 campden tablets (crushed first, then dissolved with a little cold water). Seal with rubber bungs and fit with airlocks. Move bottles to a cool and dark location with the temperature maintained at around 23°C.
9. Let sit for 1 month, siphon and rack to new bottles along with gelatin powder added to clarify.
10. Let sit for 10 days, siphon and rack to new bottles, then add 3 campden tables (crushed and dissolved in a little cold water).
11. Let sit for 6 months in a cool and dark location with the temperature around 23°C.
12. Siphon and rack to new bottles along with stabilizer added.
13. Bottle and continue aging for 2 months.

蜂蜜酒

Mead就是蜂蜜酒，石器時代的洞穴，就有人類採集蜂蜜的壁畫，歐洲古老的習俗，剛結婚的新人，整個月都喝蜜酒慶祝，這樣才能白頭偕老，所以有了「蜜月」這個名詞。基本蜜酒的原料只需要蜂蜜和水，就像葡萄酒只要有葡萄就夠了，近代釀酒科學講求品質重於一切，蜂蜜酒的釀製也不再遵循古法了。

Mead

Mead wine is made from honey. There are Stone Age paintings on cave walls that show humans collecting honey. An old tradition in Europe says that newlywed couples have to drink mead for the first month of their marriage, so that they can grow old together. That is where the word "honey moon (month)" comes from. The main ingredients for this wine are only honey and water, just as in grape wine where grapes are the only ingredient.

Mead Wine 蜂蜜酒

■ 製成份量：約4公升
■ 完成時間：2個半月可裝瓶

台灣的龍眼花蜜釀製而成的蜂蜜酒，擁有金黃透明的色澤，香醇可口，一口接一口，令人捨不得放下酒杯。

Taiwan's honey has a special sweet flavor from logan flower. Wine made with honey, has a clear golden tone, full of flowery nose. What a delightfultreat!

材料1：（第一次發酵）
※4公升蜂蜜酒的份量

蜂蜜	2公斤
熱開水	4.5公升
酵母營養劑	1小匙
單寧	1/2小匙
增酸劑	5小匙
二氧化硫	2片
白葡萄酒酵母菌	1包（5公克）
溫水 1杯	

材料2：（過程中添加）

吉利丁	2公克
二氧化硫	4片
安定劑	50c.c.

Ingredients 1（The primary fermentation）
2 liters honey
4.5 liters hot water
1 tsp. yeast nutrients
1/2 tsp. tannin
5 tsp. acid blend
1 campden tablets
1 pack green grapes wine yeast(5 grams)
1 cup lukewarm water

Ingredients 2（Add during the process）
2 grams gelatin
4 tablets sulfur dioxide
50c.c. stabilizer

■ Finished Amount : about 4 liters
■ Finishing Time : 2 1/2 months before bottling

蜂蜜

白葡萄酒酵母

二氧化硫

增酸劑

單寧粉

果膠酵素

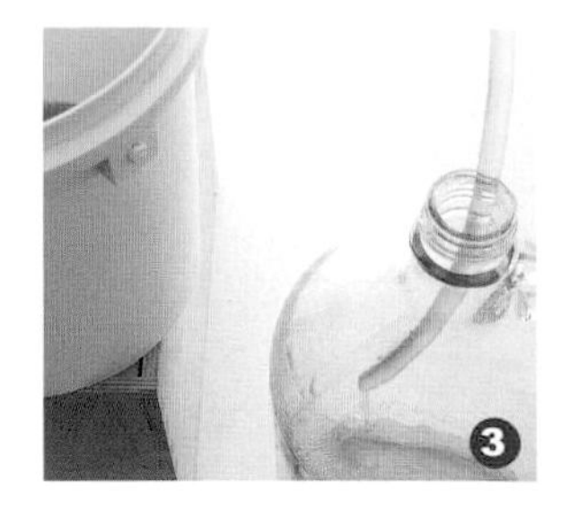

❶ 添加果膠酵素、酵母營養劑
❷ 添加單寧
❸ 將酒醪虹吸入第二次發酵瓶

做法

第一次發酵

1. 2公斤蜂蜜加入熱開水、酵母營養劑、單寧、增酸劑、二氧化硫等，倒入第一次發酵桶，放置1個小時。酵母加入1杯溫水（38˚C），10～30分鐘後倒入發酵桶，蓋好，放置陰暗處，溫度不要超過30˚C。
2. 待24小時後檢查，見氣泡自桶底昇到表面，表示發酵作用已經開始；若沒有發酵現象，再等待24小時；仍沒有動靜，用溫水加入1包酵母及1/2匙酵母營養劑，倒入發酵桶。
3. 每天用長柄匙，早晚各攪拌一次。

Steps

The primary fermentation

1. Add 2 liters of honey along with hot water, yeast nutrients, tannin, acid blend and campden tablets into the fermentor for the primary fermentation. Let sit for 1 hour. Rehydrate yeast with 1 cup of lukewarm water (38˚C) and sit for 10 to 30 minutes, then pour into the fermentor. Cover with the lid and move the fermentor to cool and dark location with the temperature maintained at less than 30˚C.
2. Let sit for 24 hours, then open to examine. If foam has formed on the surface, fermentation has started. If not, wait for another day and check again. If there is still no foam on the surface, add warm water with 1 pack of yeast and 1/2 spoonful of yeast nutrients added.
3. Stir twice a day, once in the morning and again at night, with a long stirring paddle.

第二次發酵

4. 5天後，虹吸換瓶，用橡皮塞封口，加上發酵鎖。
5. 10天後，虹吸換瓶，加入2片二氧化硫（先壓碎，以少許冷水溶解）。
6. 靜置3個月，虹吸換瓶，加入吉利丁澄清。
7. 10天後，虹吸換瓶加入2片二氧化硫（先壓碎，以少許冷水溶解）。
8. 靜置1個月，虹吸換瓶，加入安定劑。
9. 裝瓶後，繼續熟成6個月。

The secondary fermentation

4. Let sit for 5 days, siphon and rack to new bottles. Seal with rubber bungs and fit with airlocks.
5. Let sit for 5 days, siphon and rack to new bottles with 2 campden tablets added (crushed first, then dissolved in a little cold water).
6. Let sit for 3 months, siphon and rack to new bottles with gelatin added to clarify.
7. Let sit for 10 days, siphon and rack to new bottles with 2 campden tablets added (crushed first, then dissolved in a little cold water).
8. Let sit for 1 month, siphon and rack to new bottles along with stabilizer adder.
9. Bottle and continue aging for 6 months.

乾果釀酒

用乾果釀酒，比較難以掌握它的糖度，因為乾果含有大約60%的糖，會慢慢溶出到酒醪。第一次發酵完成後，用細網過濾，果渣丟棄，乾果要先切碎、去核，挑出發霉與不新鮮的。加水用果汁機打爛成果漿，充份將乾果的香甜溶到酒醪裡。

Dried Fruit Wine

It is more difficult to control the sugar gravity if dried fruit is used for making wine because dried fruit contains about 60% sugar, which will dissolve in the must very slowly. After the primary fermentation, strain with a fine strainer and remove the solids and any tiny particles. The dried fruit must be chopped and crushed very finely, any pits and spoiled pieces of the fruit must be discarded. Blend in blender with water added until well-mashed, then the sweetness of the dried fruit will be released completely in the must.

Red Date Wine
紅棗酒

■ 製成份量：約20公升
■ 完成時間：4個半月可裝瓶

常久以來，紅棗一直是泡藥酒的好藥材，其實用紅棗發酵，做成釀製酒也非常好喝。

For untold centuries red dates have been considered the best ingredients for soaking Chinese medicine. In fact, red date wine is itself delicious.

材料1：（第一次發酵）

紅棗	5公斤
糖	3公斤
柳橙	3粒
增酸劑	15小匙
熱開水	6公升
冷水	9公升
酵母營養劑	2小匙
單寧	2小匙
果膠酵素	1小匙
二氧化硫	6片
水果酒酵母	1包（5公克）
溫水	1杯

材料2：（過程中添加）

果糖	500c.c.
吉利丁	5公克
二氧化硫	3片

Ingredients 1（The primary fermentation）

5 kgs. red dates
3 kgs. sugar
3 oranges
15 tsp. acid blend
6 liters hot boiling water
9 liters cold water
2 tsp. yeast nutrients
2 tsp. tannin
1 tsp. pectin enzyme
6 campden tablets(crushed)
1 pack fruit wine yeast(5 grams)
1 cup lukewarm water

Ingredients 2（Add during the process）

500c.c. fructose
5 grams gelatin
3 campden tablets(crushed)

■ Finished Amount : about 20 liters
■ FInishing Time : 4 1/2 months before bottling

紅棗

柳橙

二氧化硫

水果酒酵母

糖

單寧粉

果膠酵素

❶ 紅棗及切碎原料
❷ 使用去核切碎的紅棗，加入果汁機打碎
❸ 過濾除渣

做法

第一次發酵

1. 紅棗去核，分數次用果汁機打碎，柳橙切片，除了酵母，全部原料放入第一次發酵桶，攪拌混合，放置1小時。
2. 酵母加入1杯溫水，10～30分鐘後，倒入發酵桶，用力攪拌，蓋好發酵桶，放在23℃左右陰暗處。
3. 待24小時後檢查，見果肉及泡沫浮在表面，表示發酵作用已經開始；若沒有發酵現象，再等待24小時；仍沒有動靜，請參考P.111（釀酒Q&A）問題1的原因解答。
4. 每天用長柄匙，早晚各攪拌一次。

Steps

The primary fermentation

1. Remove pits from red dates and blend in blender little at a time until well mashed. Slice oranges into thin pieces. Place all the ingredients (except yeast) in the fermentor for the primary fermentation. Stir until well-mixed and sit for 1 hour.
2. Rehydrate yeast in 1 cup of lukewarm water for 10 to 30 minutes. Transfer to the fermentor and stir vigorously until even. Cover with the lid and move the fermentor to a cool and dark location with the temperature maintained at around 23℃.
3. Let sit for 24 hours, then open to examine. If foam has formed on the surface, fermentation has started. If not, wait for another day and check again. If there is still no foam on the surface, refer to Question & Answer 1 on p.111〔Wine Making Q&A〕.
4. Stir twice a day, once in the morning and again at night, with a long stirring paddle.

第二次發酵

5. 每隔1天，用比重測糖器檢測糖度。當含糖量降到5度以下時，用細網過濾，擠乾酒醪，果渣丟棄，換入第二次發酵桶，用橡皮塞封口，加上發酵鎖，觀察到發酵作用停止時（發酵鎖不再有氣泡冒出）。

6. 加入果糖，搖動發酵桶，讓酒醪和果糖混合。仍用橡皮塞封口，加上發酵鎖，放置在23℃陰暗處，發酵作用再次開始，等到發酵作用停止後，虹吸換瓶，加入吉利丁澄清。

7. 10天後，虹吸換瓶，加入3片二氧化硫（先壓碎，以少許冷水溶解）。

8. 靜置3個月。

9. 裝瓶後，繼續熟成6個月。

The secondary fermentation

5. Measure the sugar gravity of the must with a hydrometer every other day. If the sugar gravity falls below 5, strain with a fine strainer, squeeze out the excess liquid from the solids and discard. Rack to another fermentor for the secondary fermentation. Seal with a rubber bung and fit with airlock. Observe until the fermentation has stopped (when there is no bubbles coming out from the airlock).

6. Add fructose and shake the fermentor to mix the must and fructose well. Seal with rubber cork and fit with airlock again. Remove to a cool corner with the temperature maintained at around 23°C. The fermentation will start all over again, wait until the fermentation has stopped, siphon and rack to bottles along with gelatin added to clarify.

7. Let sit for 10 days, siphon and rack to new bottles along with 3 campden tablets added (crush first, then dissolve in a little cold water).

8. Let sit for 3 months.

9. Bottle and continue aging for 6 months.

Dried Logan Wine

龍眼乾酒

■ 製成份量：約18公升
■ 完成時間：5個月可裝瓶

龍眼乾加蜂蜜，等於茶色香醇的甜酒，寒流來襲的夜晚，只要喝一小杯就能暖到心底。

Logan with honey wine is brown, fragrant and sweet. On chilly nights, one small drink will warm the corners of your heart.

材料1：（第一次發酵）

龍眼乾	5公斤
蜂蜜	2公斤
葡萄柚汁	500c.c.
增酸劑	10小匙
酵母營養劑	2小匙
單寧	2小匙
熱開水	6公升
冷水	8公升
二氧化硫	6片
水果酒酵母	1包（5公克）
溫水	1杯

材料2：（過程中添加）

果糖	400c.c.
吉利丁	5公克
二氧化硫	3片

Ingredients 1（The primary fermentation）

5 kgs. dried logan
2 kgs. honey
500c.c. grapefruit juice
10tsp. acid blend
2 tsp. yeast nutrients
2 tsp. tannin
6 liters hot water
8 liters cold water
6 campden tablets
1 pack fruit wine yeast(5 grams)
1 cup warm water

Ingredients 2（Add during the process）

400c.c. fructose
5 grams gelatin
3 campden tablets

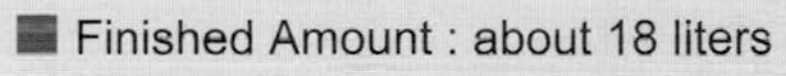

■ Finished Amount : about 18 liters
■ Finishing Time : 5 months before bottling

龍眼

蜂蜜

二氧化硫

水果酒酵母

葡萄柚汁

單寧粉

果膠酵素

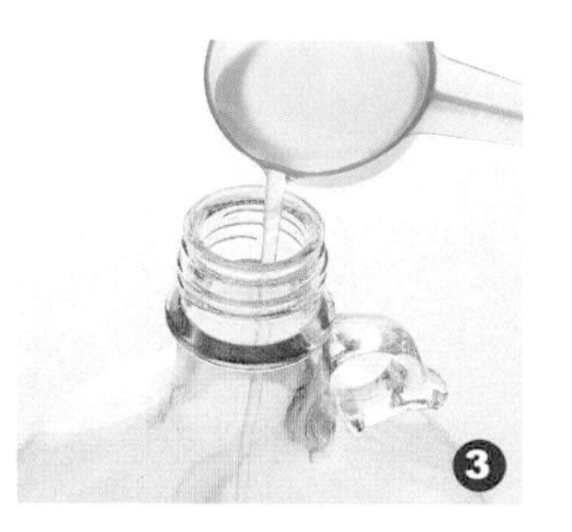

❶ 桂圓肉加水打碎
❷ 用豆漿袋過濾
❸ 添加吉利丁

做法

第一次發酵

1. 龍眼乾加水，分數次用果汁機打碎，除了酵母，所有原料全部倒入第一次發酵桶，攪拌混合，放置1小時，發酵桶要蓋好。
2. 酵母加入1杯溫水（38°C），10～30分鐘，倒入發酵桶，攪拌，蓋好，放在23°C左右陰暗處。
3. 待24小時後檢查，見果肉及泡沫浮在表面，表示發酵作用已經開始；若沒有發酵現象，再等待24小時；仍沒有動靜，請參考P.111〈釀酒Q&A〉問題1的原因解答。
4. 每天用長柄匙，早晚各攪拌一次。

Steps

The primary fermentation

1. Add water to dried logan and blend in blender little at a time until all is well mashed. Add all the ingredients (except yeast) to the fermentor for the primary fermentation. Stir well and let sit for one hour. Cover tightly with the lid of the fermentor.
2. Rehydrate yeast with 1 cup of lukewarm water (38°C) for 10 to 30 minutes. Pour into the fermentor and mix well. Put the cover back and move the fermentor to a cool and dark location with the temperature maintained at around 23°C.
3. Let sit for 24 hours, then open to examine. If foam has formed on the surface, fermentation has started. If not, wait for another day and check again. If there is still no foam on the surface, refer to Question & Answer 1 on p.111〔Wine Making Q&A〕.
4. Stir twice a day, once in the morning and again at night, with a long stirring paddle.

第二次發酵

5. 每隔1天，用比重測糖器檢測糖度。當含糖量降到5度以下時，用豆漿袋或細網過濾，擠乾酒醪，果渣丟棄。
6. 酒醪換入第二次發酵果酒瓶，用橡皮塞封口，加上發酵鎖。
7. 3～7天左右，發酵鎖不再有氣泡冒出，表示發酵作用已經停止，加入果糖搖動果酒瓶，讓酒醪和果糖充份混合，用橡皮塞封口，加上發酵鎖。
8. 發酵作用再次開始，當觀察到發酵作用停止時，虹吸換瓶，加入3片二氧化硫（先壓碎，以少許冷水溶解）。
9. 靜置3個月。
10. 虹吸換瓶，加入吉利丁澄清劑。
11. 10天後虹吸換瓶，加入安定劑。
12. 靜置1個月，裝瓶。
13. 裝瓶後，繼續熟成9個月。

The secondary fermentation

5. Measure the sugar gravity of the must every other day with a hydrometer. When the sugar gravity is below 5, strain with a soybean bag or fine strainer, squeeze the excess liquid out of the solids and discard.
6. Rack the must to new bottles for the secondary fermentation. Seal with rubber bungs and fit with airlocks.
7. Let sit for 3 to 7 days until there is no foam coming out of the airlock, which means that the fermentation has already stopped. Add fructose and shake the bottles well to thoroughly combine the must and fructose. Seal with rubber bungs and fit with airlocks again.
8. When fermentation starts once again, observe until the fermentation stops, siphon and rack to new bottles along with 3 campden tablets added (crushed first, then dissolved in a little water).
9. Let sit for 3 months.
10. Siphon and rack to new bottles with gelatin added to clarify.
11. Let sit for 10 days, siphon and rack to new bottles along with stabilizer added.
12. Let sit for 1 month, and bottle.
13. Continue aging for 9 months.

Raisin Wine
葡萄乾酒

■ 製成份量：約20公升
■ 完成時間：5個月可裝瓶

琥珀色的葡萄乾酒包容性很強，當你的釀酒技術熟練後，可以隨季節加入可食用鮮花或香草，創造你自己的獨家秘方。

A wide range of ingredients can be incorporated into the amber depths of raisin wine. Experiment by adding edible flowers or herbs to the wine and create your own unique recipe.

材料1：（第一次發酵）

紅葡萄乾	6公斤
糖	4公斤
增酸劑	16小匙
熱開水	6公升
酵母營養劑	2小匙
單寧	3小匙
果膠酵素	2小匙
二氧化硫	6片
冷水	8公升
紅葡萄酒酵母	1包（5公克）
溫水	1杯

材料2：（過程中添加）

吉利丁粉	5公克
二氧化硫片	3片
安定劑	250c.c.

Ingredients 1 (The primary fermentation)

6 kgs red raisins
4 sugar
16 tsp. acid blend
6 liters hot water
2 tsp. yeast nutrients
3 tsp. tannin
2 tsp. pectin enzyme
6 campden tablets
8 liters cold water
1 pack red wine yeast(5 grams)
1 cup lukewarm water

Ingredients 2 (Add during the process)

5 grams gelatin powder
3 campden tablets
250c.c. stabilizer

■ Finished Amount : about 20 liters
■ Finishing Time : 5 months before bottling

葡萄乾

紅葡萄酒酵母

二氧化硫

糖

單寧粉

果膠酵素

❶ 葡萄乾加水打碎
❷ 虹吸換瓶
❸ 已澄清的葡萄乾酒

做法

第一次發酵

1. 葡萄乾切碎，開水將糖溶解，全部原料（除了酵母），倒入第一次發酵桶，攪拌，放置1小時。
2. 酵母放入溫水（38℃）中，10～30分鐘，倒入發酵桶，用力攪拌，蓋好，放在23℃左右陰暗處。
3. 待24小時後檢查，見果肉及泡沫浮在表面，表示發酵作用已經開始；若沒有發酵現象，再等待24小時；仍沒有動靜，請參考P.111（釀酒Q&A）問題1的原因解答。
4. 每天用長柄匙，早晚各攪拌一次。

Steps

The primary fermentation

1. Chop the raisins finely. Meanwhile, thoroughly dissolve the sugar in hot water. Add all the ingredients (except yeast) to the fermentor for the primary fermentation. Stir well and let sit for 1 hour.
2. Rehydrate yeast with lukewarm water (38℃) for 10 to 30 minutes. Pour into the fermentor and stir vigorously until mixed. Cover and move to a dark and cool location with the temperature maintained at around 23℃.
3. Let sit for 24 hours, then open to examine. If foam has formed on the surface, fermentation has started. If not, wait for another day and check again. If there is still no foam on the surface, refer to Question & Answer 1 on p.111〔Wine Making Q&A〕.
4. Stir twice a day, once in the morning and again at night, with a long stirring paddle.

W

winemaking at home

第二次發酵

5. 每隔1天，用比重測糖器檢測糖度。當含糖量降到5度以下時，用細網過濾，擠乾酒醪，丟棄果渣。

6. 將酒醪換入第二次發酵用的果酒瓶，用橡皮塞封口，加發酵鎖。

7. 10天後虹吸換瓶。

8. 靜置3週，虹吸換瓶，加入澄清劑。

9. 10天後虹吸換瓶加入3片二氧化硫（先壓碎，以少許冷水溶解）。

10. 靜置3個月。

11. 虹吸換瓶，加入安定劑。

12. 裝瓶後，繼續熟成3個月。

The secondary fermentation

5. Measure the sugar gravity of the must every other day with a hydrometer. When the sugar gravity is below 5, strain with a fine strainer, squeeze the excess liquid out of the solids and discard.

6. Rack must to new bottles for the second fermentation. Seal with rubber bungs and fit with airlocks.

7. Let sit for 10 days, siphon and rack to new bottles.

8. Let sit for 3 weeks, siphon and rack to new bottles along with clearing agent.

9. Let sit for 10 days, siphon and rack to new bottles along with 3 campden tablets added (crushed first, then dissolved in a little cold water).

10. Let sit for 3 months.

11. Siphon and rack to bottles along with stabilizer added. Bottle.

12. Continue aging for 3 months.

Almond Wine
杏仁酒

■ 製成份量：約4公升
■ 完成時間：4個月可裝瓶

材料1：（第一次發酵）

杏仁片	100公克
葡萄乾	500公克
白葡萄濃縮汁 (可以新鮮青葡萄去皮榨汁取代)	250c.c.
糖	1公斤
檸檬汁	1/2杯
檸檬皮碎	1小匙
酵母營養劑	1小匙
果膠酵素	1小匙
單寧	1小匙
二氧化硫	1片
冷水	3公升
水果酒酵母	1包（5公克）
溫水	1杯

材料2：（過程中添加）

吉利丁	2公克
二氧化硫	1片
安定劑	80c.c.

淡淡的杏仁香，配上淡淡的酒香，下午茶時間，搭配一碟蛋糕或是水果派，人生多麼美好啊！

A light almond fragrance and a light wine bouquet with a plate of cake or fruit pie for afternoon tea time-- what a beautiful life that is!

Ingredients 1（The primary fermentation）

100 grams almond flakes
500 grams raisins
250c.c. green grapes juice concentrate
1 liter sugar
1/2 cup lemon juice
1 tsp. lemon zest
1 tsp. yeast nutrients
1 tsp. pectin enzyme
1 tsp. tannin
1 campden tablets
3 liters cold water
1 pack fruit wine yeast(5 grams)
1 cup lukewarm water

Ingredients 2（Add during the process）

2 grams gelatin
1 campden tablets
80c.c. stabilizer

■ Finished Amount : about 4 liters
■ Finishing Time : 4 months before bottling

杏仁碎

檸檬汁

檸檬皮

二氧化硫

白葡萄濃縮汁

單寧粉

果膠酵素

❶ 虹吸換瓶
❷ 添加吉利丁
❸ 添加二氧化硫

做法

第一次發酵

1. 杏仁和葡萄乾切碎，放入鍋中，加3公升水，大火煮滾後轉小火燜煮1小時，要隨時加水，保持3公升的份量。
2. 將煮好的汁液以細網過濾，放入第一次發酵桶，加入葡萄濃縮汁、糖、檸檬汁、檸檬皮碎、酵母營養劑、單寧、果膠酵素、二氧化硫及冷水攪拌，放置1小時。
3. 酵母加入1杯溫水中，10～30分鐘後，倒入發酵桶，攪拌蓋好，放在23°C左右陰暗處。
4. 待24小時後檢查，見果粒及泡沫浮在表面，表示發酵作用已經開始；若沒有發酵現象，再等24小時；仍沒有動靜，請參考P.111（釀酒Q&A）問題1的原因解答。
5. 每天用長柄匙，早晚各攪拌一次。

Steps

The primary fermentation

1. Chop almonds and raisins together finely. Transfer to boiling pot with 3 liters of water. Bring to a boil over high heat and then reduce heat to low. Simmer for 1 hour, add water constantly to maintain a 3 liter portion.
2. Strain the solution with a fine strainer and transfer to the fermentor for the primary fermentation. Add grape juice concentrate, sugar, lemon juice, lemon zest, yeast nutrients, tannin, pectin enzyme, campden tablets and cold water. Stir vigorously until well-mixed. Let sit for 1 hour.
3. Rehydrate yeast with 1 cup of lukewarm water for 10 to 30 minutes. Pour into the fermentor and cover well. Remove to a cool and dark location with the temperature maintained at around 23°C.
4. Let sit for 24 hours, then open to examine. If foam has formed on the surface, fermentation has started. If not, wait for another day and check again. If there is still no foam on the surface, refer to Question & Answer 1 on p.111〔Wine Making Q&A〕.
5. Stir twice a day, once in the morning and again at night, with a long stirring paddle.

W

winemaking at home

第二次發酵

6. 每隔1天，用比重測糖器檢測果糖含量。當糖含量降到5度以下時，虹吸換瓶使用4公升果酒瓶，用橡皮塞封口，加發酵鎖。
7. 10天後，虹吸換瓶，加入吉利丁粉澄清。
8. 10天後，虹吸換瓶，加入1片二氧化硫（先壓碎，以少許冷水溶解）。
9. 靜置3個月。
10. 虹吸加入安定劑，裝瓶。
11. 裝瓶後，繼續熟成3個月。

The secondary formentation

6. Measure the sugar gravity of the must every other day with a hydrometer. If the sugar gravity is below 5, siphon and rack to 4-liter fruit bottles. Seal with rubber bungs and fit with airlocks.
7. Sit for 10 days, siphon and rack to new bottles along with gelatin powder added to clear the must.
8. Let sit for 10 days, siphon and rack to new bottles along with 1 campden tablet (crushed first, then dissolved in a little cold water).
9. Let sit 3 months.
10. Siphon and add stabilizer, then bottle.
11. Continue aging for 3 more months.

Garlic Wine 大蒜酒

■ 製成份量：約4公升
■ 完成時間：4個月可裝瓶

食譜中使用的大蒜沒有切碎，不會有沖鼻的大蒜味，用大蒜酒調製沙拉醬或醃泡肉類，給你的味蕾帶來全新的驚喜。

The garlic in this recipe is not crushed, so there will not be a strong garlic flavor. Use garlic wine to make salad dressings or in marinating meat for a new taste surprise.

材料1：（第一次發酵）

整球大蒜	12個
白葡萄濃縮汁	600c.c.
(可以新鮮青葡萄去皮榨汁取代)	
檸檬汁	100c.c.
檸檬皮碎	2小匙
酵母營養劑	1小匙
溫水	2.5公升
糖	1公斤
果膠酵素	1小匙
單寧	1/2小匙
二氧化硫片	1片
水果酒酵母	1包（5公克）
溫水	1杯

材料2：（過程中添加）

二氧化硫	1片
安定劑	100c.c.

Ingredients 1 (The primary fermentation)

12 heads garlic
600c.c. green grapes juice concentrate
100c.c. lemon juice
2 tsp. lemon zest
1 tsp. yeast nutrients
1.5 liters lukewarm water
1 liter sugar
1 tsp. pectin enzyme
1/2 tsp. tannin
1 campden tablet
1 pack fruit wine yeast(5 grams)
1 cup lukewarm water

Ingredients 2 (Add during the process)

1 campden tablet
100c.c. stabilizer

■ Finished Amount : about 4 liters
■ Finishing Time : 4 months before bottling

烤過之大蒜

檸檬汁

檸檬皮

二氧化硫

白葡萄濃縮汁

單寧粉

果膠酵素

❶ 生大蒜及烤好的大蒜下鍋同煮
❷ 煮好之大蒜，過濾汁液
❸ 添加二氧化硫

做法

第一次發酵

1. 12球大蒜分成8球及4球兩份，剝去外皮成蒜瓣，4球的蒜瓣用鋁箔紙包好，放入350℃烤箱烤2小時，讓大蒜中的糖焦化。
2. 將烤好的蒜瓣及8球新鮮蒜瓣，放入鍋中，不加油、加2公升水，大火煮滾後轉小火燜煮40分鐘，要隨時加水，保持2公升的份量。關火時加入檸檬皮碎，繼續燜1小時。
3. 用細網過濾，丟棄殘渣，將煮好的汁液放入第一次發酵桶，加入白葡萄濃縮汁、檸檬汁、酵母營養劑、溫水、糖及果膠酵素、單寧、二氧化硫片攪拌，放置1小時。
4. 酵母加入1杯溫水中，10～30分鐘後，倒入發酵桶攪拌，放在陰暗處。
5. 待24小時後檢查，見氣泡自桶底冒到表面，表示發酵作用已經開始；若沒有發酵現象，再等待24小時；仍沒有動靜，重覆步驟4的動作。

Steps

The primary fermentation

1. Divide 12 heads of garlic into two groups, 8 and 4. Peel off the skin and separate the cloves. Wrap the garlic cloves from 4 heads with aluminum foil and bake in the oven at 35℃ for 2 hours to caramelize the sugar in the garlic.
2. Place the baked garlic cloves along with the 8 fresh garlic cloves in the boiling pot. Add 2 liters of water and bring to a boil over high heat. Then reduce heat to low and simmer for 40 minutes. Add water constantly to maintain the 2 liter level. Add the lemon zest when removing from heat. Continue to simmer for 1 hour.
3. Strain with a fine strainer, remove the solids. Transfer the garlic solution to the fermentor for the primary fermentation. Add white grape juice concentrate, lemon juice, yeast nutrients, lukewarm water, sugar as well as pectin enzyme, tannin and campden tablets. Let sit for 1 hour.
4. Rehydrate yeast with 1 cup of lukewarm water. Let sit for 10 to 30 minutes to dissolve. Pour into the fermentor and mix well. Move the fermentor to a cool and dark location..
5. Let sit for 24 hours, then open to examine. If foam has formed on the surface, fermentation has started. If not, wait for another day and check again. If there is still no foam on the surface, wait for another 24 hours and check again. If there is still no foam, repeat step 4.

◎釀酒TIPS
大蒜酒屬於料理酒，所以不必澄清。

Winemaking Tips
Garlic wine is a cooking wine and does not need to be clarified.

W
winemaking at home

第二次發酵

6. 10天後虹吸換瓶，用4公升果酒瓶，橡皮塞封口，加發酵鎖。
7. 20天後，虹吸過濾，加入二氧化硫（先壓碎，以少許冷水溶解）。
8. 靜置3個月。
9. 加入安定劑，裝瓶。
10. 繼續熟成3個月。
11. 大蒜酒、橄欖油及紅酒醋以同比例放入玻璃瓶，加入任何你喜歡的香草（羅勒、迷迭香等），就是最美味的沙拉醬。用大蒜酒加調味料醃肉，絕對是一定要嘗試的好滋味。

The secondary fermentation

6. Let sit for 10 days, siphon and rack to 4-liter glass wine bottle. Seal with rubber bungs and fit with air-locks.
7. Let sit for 20 days, slphon and rack bottles with campden tablets added (crushed first, then dissolved in a little cold water).
8. Let sit for 3 months.
9. Add stabilizer and bottle.
10. Continue aging for 3 more months.
11. With the same portion of garlic wine, olive oil and red wine vinegar in the glass jar, add herbs as desired (basil, rosemary, etc) to make a delicious salad dressing. Also, try using the garlic wine to marinate meat. The flavor is wonderful.

濃縮葡萄汁釀酒

台灣種植的葡萄有兩大類：釀酒葡萄和鮮食葡萄，釀酒葡萄以黑后和金香兩個品種最普遍。鮮食葡萄也是釀酒的好原料。只是它的價格比較貴，台灣氣候溫暖，葡萄種植技術精良，每年收穫兩次的葡萄，使得種植葡萄獲利更好。相對的，葡萄樹沒有冬眠，休養生機的機會，葡萄自開花到結果也沒有足夠的時間孕育它的成熟風味。台灣的葡萄可以釀出美味的葡萄酒，但不是頂極的葡萄酒。

本節將介紹幾種不同葡萄品種濃縮汁釀製而成的葡萄酒。葡萄濃縮汁有很多不同的品牌，簡單的分類只有兩種：

1. 單一釀酒葡萄品種濃縮汁：純的葡萄汁濃縮，低溫抽取百分之70的水份，添加適量的二氧化硫增加保存期限，糖度68～70゜。
2. 混合品種葡萄汁：使用的葡萄品種不詳，甚至有的品牌還混雜其它水果在內，降低成本，濃縮時添加糖、檸檬酸等，糖度30～77゜。

Grape Juice Concentrate Wines

There are two kinds of grapes in Taiwan, one for making wine and one for eating. Grapes used in winemaking are usually Hei-hou and Chin-hsiang. Grapes for eating are also good for making wine, but the price is much higher. Taiwan has a warm climate that is good for growing grapes, and local growing techniques are precise and advanced. Grapes are produced twice a year and farmers who grow grapes make good money. On the other hand, grape vines do not go dormant, so there is no time for breaking and resting. Grapes do not get enough time from blooming to harvesting ripen properly, thus, grapes in Taiwan can be made into wine, but not truly first-class wine.

French Bordeaux Red Wine

法國波爾多式葡萄紅酒

■ 製成份量：約20公升
■ 完成時間：4個月可裝瓶

Cabernet Sauvignon葡萄是世界上種植最多的品種，法國波爾多紅葡萄酒是它的傑作，顏色呈現深紫紅色、單寧高，需要熟成。喝到口中，帶有些澀又有些黑莓果的香味，口感多樣化。

The cabernet sauvignon grape is a common variety. The Bordeaux from France is its masterpiece. The color is a deep purple-red, high in tannins, and requires aging. It is a little dry upon initial taste, and sometimes tastes like blackberries due to its many flavors.

材料1：（第一次發酵）

Cabernet Sauvignon濃縮汁	3公升
Merlot濃縮汁	1公升
糖	1.5公斤
熱開水	6公升
增酸劑	4小匙
酵母營養劑	1小匙
單寧	2小匙
冷水	9公升
紅葡萄酒酵母	1包（5公克）
溫水	1杯

材料2：（過程中添加）

吉利丁粉	5公克
二氧化硫	3片
安定劑	100c.c.
食用精製甘油	30c.c.

Ingredients 1 (The primary fermentation)

3 liters Cabernet Sauvignon concentrate
1 liter Merlot concentrate
1.5 liters sugar
6 liters hot water
4 tsp. acid blend
1 tsp. yeast nutrients
2 tsp. tannin
9 liters cold water
1 pack red wine yeast(5 grams)
1 cup lukewarm water

Ingredients 2 (Add during the process)

5 grams gelatin powder
3 campden tablets
100c.c. stabilizer
30c.c. glycerin

■ Finished Amount : about 20 liters
■ Finishing Time : 4 months before bottling

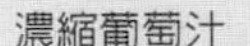
濃縮葡萄汁

紅葡萄酒酵母

二氧化硫

糖

單寧粉

增酸劑

❶ 濃縮葡萄汁加入第一次發酵桶中
❷ 正在發酵中
❸ 裝瓶前添加安定劑

做法

第一次發酵

1 . 糖溶於熱開水，葡萄濃縮汁、增酸劑、酵母營養劑、單寧、冷水全部放入第一次發酵桶。
2 . 酵母加入1杯溫水（38℃）10～30分鐘，倒入發酵桶蓋好，放在23℃左右的陰暗處。
3 . 待24小時後檢查，見氣泡自桶底昇到表面，表示發酵作用已經開始；若沒有發酵現象，再等24小時；仍沒有動靜，請參考P.111（釀酒Q&A）問題1的原因解答。
4 . 每天用長柄匙，早晚各攪拌一次。

Steps

The primary fermentation

1. Dissolve sugar in hot water, then pour into the fermentor along with grape juice concentrate, acid blend, yeast nutrients, tannin and cold water for the primary fermentation.
2. Rehydrate yeast with 1 cup of lukewarm water (38°C) for 10 to 30 minutes. Transfer to the fermentor and cover with a lid. Then move to a cool and dark location with temperature maintained at around 23°C.
3. Let sit for 24 hours, then open to examine. If foam has formed on the surface, fermentation has started. If not, wait for another day and check again. If there is still no foam on the surface, refer to Question & Answer 1 on p.111〔Wine Making Q&A〕.
4. Stir twice a day, once in the morning and again at night, with a long stirring paddle.

◎釀酒TIPS
本道酒譜添加食用精製甘油的目的，是為了增加酒的質感。

Winemaking Tips
Glycerin is added here to enhance the body of the wine.

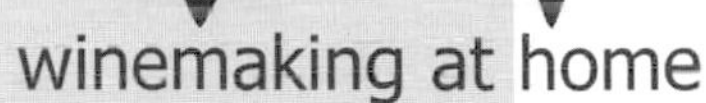

第二次發酵

5. 每隔1天，用比重測糖器檢測糖度。當含糖量降到5度以下時，虹吸到第二次發酵用的果酒瓶，用橡皮塞封口，加發酵鎖，放到20°C左右的陰暗處。

6. 10天後，虹吸換瓶，加入吉利丁澄清。

7. 10天後，虹吸換瓶，加入食用甘油、二氧化硫（先壓碎，以少許冷水溶解）。

8. 靜置3個月。

9. 虹吸換瓶加入安定劑、裝瓶。

10. 裝瓶後，繼續熟成1年。

The secondary fermentation

5. Measure the sugar gravity with a hydrometer every other day. If the sugar gravity is below 5, siphon and rack to bottles for the second fermentation. Seal with rubber bungs and fit with airlocks. Move the bottles to cool and dark location with the temperature around 20°C.

6. Let sit for 10 days, siphon and rack to new bottles along with gelatin added to clarify.

7. Let sit for 10 days, siphon and rack to new bottles along with glycerin and campden tablets added (crushed first, dissolved in a little cold water).

8. Let sit for 3 months.

9. and rack to new bottles with stabilizer added.

10. Bottle and continue aging for 1 year.

Sauvignon Blanc Wine

加州式白葡萄酒

■ 製成份量：約20公升
■ 完成時間：4個月可裝瓶

Sauvignon Blanc葡萄，帶有草原的氣息，熟成後的葡萄酒，清脆夾有酒香，很順口；最適合搭配清淡的食物。

Sauvignon Blanc is a grape, whose flavor suggests fields of grass. After aging, the wine is clear with a crisp, smooth flavor. It is best served with a light dish.

材料1：（第一次發酵）

Sauvignon Blanc濃縮汁	3公升
新鮮綠葡萄	1公斤
糖	2公斤
增酸劑	1小匙
熱開水	6公升
酵母營養劑	1小匙
果膠酵素	1小匙
單寧	2小匙
冷水	8公升
白葡萄酒酵母	1包（5公克）
溫水	1杯

材料2：（過程中添加）

皂土	2公克
二氧化硫	3片
安定劑	200c.c.

Ingredients 1（The primary fermentation）

3 liters Sauvignon Blanc
1 liter fresh green grapes
2 liters sugar
1 tsp. acid blend
6 liters hot water
1 tsp. yeast nutrients
1 tsp. pectin enzyme
2 tsp. tannin
8 liters cold water
1 pack white wine yeast(5 grams)
1 cup lukewarm water

Ingredients 2（Add during the process）

2 grams bentonite
3 campden tablets
200c.c. stabilizer

■ Finished Amount : about 20 liters
■ Finishing Time : 4 months before bottling

白葡萄

白葡萄酒酵母

二氧化硫

糖

單寧粉

果膠酵素

❶ 白葡萄酒使用之原料
❷ 添加酵母（已活物化）
❸ 添加二氧化硫

做法

第一次發酵

1. 新鮮葡萄壓碎，所有原料全都倒入第一次發酵桶，除了酵母。
2. 酵母加入1杯溫水，10～30分鐘，倒入發酵桶拌勻，蓋好，放在23℃左右的陰暗處。
3. 待24小時後檢查，見氣泡自桶底昇到表面，表示發酵作用已經開始；若沒有發酵現象，再等待24小時；仍沒有動靜，請參考P.111（釀酒Q&A）問題1的原因解答。
4. 每天用長柄匙，早晚各攪拌一次。

Steps

The primary fermentation

1. Crush the fresh grapes and pour all the ingredients (except yeast) into the fermentor for the primary fermentation.
2. Rehydrate yeast with 1 cup of lukewarm water for 10 to 30 minutes. Pour into the fermentor and mix well. Cover with a lid and move to a cool and dark location with the temperature maintained at around 23℃.
3. Let sit for 24 hours, then open to examine. If foam has formed on the surface, fermentation has started. If not, wait for another day and check again. If there is still no foam on the surface, refer to Question & Answer 1 on p.111〔Wine Making Q&A〕.
4. Stir twice a day, once in the morning and again at night, with a long stirring paddle.

W

winemaking at home

第二次發酵

5．每隔1天，用比重測糖器檢測糖度。當含糖量降到5度以下時，用細網擠乾果汁，丟棄果肉皮渣，酒醪換入第二次發酵用的果酒瓶，用橡皮塞封口，加上發酵鎖，放在18°C左右的陰暗處。
6．10天後虹吸換瓶，加入皂土澄清。
7．10天後換桶，加入3片二氧化硫（先壓碎，以少許冷水溶解）。
8．靜置3個月，裝瓶。
9．裝瓶後，繼續熟成9個月。

The secondary fermentation

5. Measure sugar gravity of the must every other day with a hydrometer. When the sugar gravity is below 5, strain the must through a strainer thoroughly and remove the solids. Rack to bottles for the secondary fermentation. Seal with rubber bungs and fit with airlocks, then move the bottles to cool and dark location with the temperature around 18°C.
6. Let sit for 10 days, siphon and rack to new bottles with bentonite added to clarify.
7. sit for 10 days, siphon and rack to new bottles along with 3 campden tablets added (crushed first, then dissolved with a little cold water).
8. Let sit for 3 months and bottle.
9. Continue aging for 9 months.

Advanced Techniqu

Chapter III

酸 Acid
糖 Sugar
酒精度 Alcohol Gravity
二氧化硫 Sulfite
溫度 Temperature
酵母 Yeast
澄清 Clearing/Clarifying
陳年 Aging
勾對 Mixing
調味 Flavor
裝瓶 Bottling
軟木塞 Soft Corks/Stoppers
軟木塞機 Cork Machine

es

3

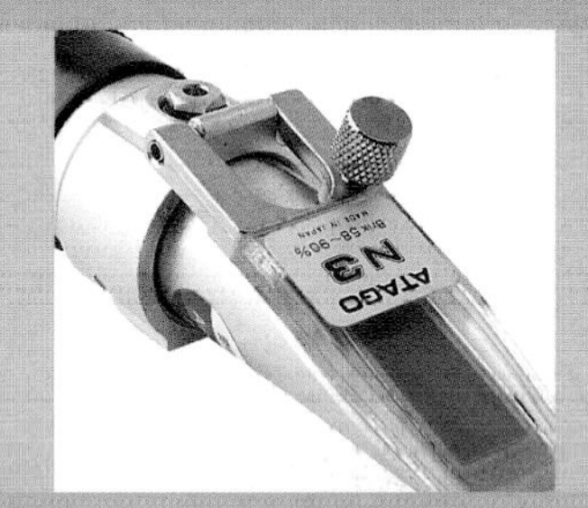

技術篇

酸 Acid
糖 Sugar
酒精度 Alcohol Gravity
二氧化硫 Sulfite
溫度 Temperature
酵母 Yeast
澄清 Clearing/Clarifying
陳年 Aging
勾對 Mixing
調味 Flavor
裝瓶 Bottling
軟木塞 Soft Corks/Stoppers
軟木塞機 Cork Machine

總論

如果你翻到了這一篇，表示你對釀酒已經真正的產生了興趣，希望能多知道一些有關釀酒的技術。的確，釀酒的過程是很科學化的，酵母菌、糖、酸等物質經過發酵作用，將單純的果汁轉變成香醇濃郁的美酒，目前科學家已經發現葡萄酒中含有兩百多種芳香物質及揮發性酯類。也難怪酒是那樣的千變萬化，讓人著迷。

書中你常會看到「酒醪」這個專有名詞，指的是釀酒原料經過壓碎、搾汁、切碎、煮爛等程序，準備成適合釀酒的狀態，酒醪的含糖量及酸度，就是釀酒成功與否的關鍵。接下來我們把酒醪轉變成含有酒精濃度的飲料之間的幾個技術層面仔細說明，當你具備了有關這方面足夠的知識，你就能輕而易舉研發出適合自己口味的酒譜了。

Conclusion

This section of the book introduces the techniques integral to winemaking. The process of making wine is very scientific. Simple materials such as yeast, sugar, and acid transform a simple liquid into a fragrant, delicious, and thick wine. Scientists have discovered that grape wine contains over two hundred kinds of aromatic materials and esters. No wonder wine is so charming and comes in so many varieties.

In this book the special term "must" often appears. This refers to the ingredients after the process of crushing, squeezing, chopping, and boiling fruits in preparation for winemaking. The sugar gravity and acid gravity of the must are the keys to the success of the wine. Next we must discuss in detail a few skills which transform the must into a beverage which contains alcohol. When you have sufficient knowledge and understanding of this process, you will be able to create your wine recipes to suit your tastes.

釀酒成功的要訣

1.酸：酒中酸度不夠，喝到口中會有苦味且平淡無味；若是含酸太多，則會酸澀難以入喉。釀酒時將原料中所含的酸調整到平衡點是很重要的。

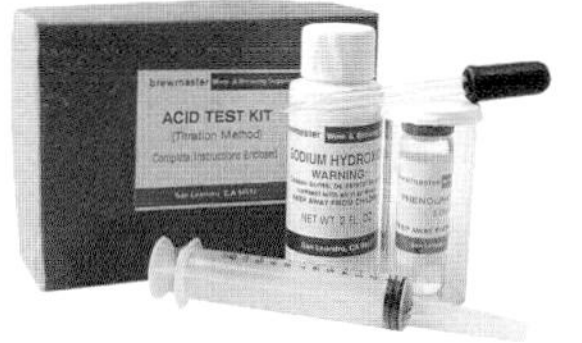

整套測酸工具

釀酒前一定要先檢測果汁或酒醪中的含酸量及含糖量，檢測含酸量最常用的方法是用滴酸儀器。仔細取出你需要的果汁或酒醪（若含有果肉渣，需要先過濾，被取澄清汁）將儀器裡的鹼液一滴一滴加到酸性果汁中，觀查顏色的變化，當果汁或酒醪因鹼液的滴入而中和時，計算滴入了多少的鹼液，就可以得到酒醪的酸度，當你購買滴酸儀器時，盒內均附有詳細的使用說明。

淺顏色的果汁或酒醪很容易從顏色的變化上，觀察到酸鹼是否已得到中和。紅葡萄汁因為是深紫紅的顏色，用肉眼觀測比較難以抓到酸鹼中和的正確時機，此時有兩個補救方法：一個是在紅葡萄汁中加入蒸餾水，沖淡它的顏色，因為蒸餾水是中性，並不會增加或減少紅葡萄汁的酸度；另一個方法是使用試紙，當覺得滴入的鹼液差不多夠了，這時滴入一點已加鹼液的果汁或酒醪在試紙上，觀查試紙顏色的變化。試紙呈紅色表示酸性，尚未達到中和；試紙呈藍色表示鹼性，則表示鹼液加多了。

釀酒時最理想的酸度是多少呢？紅葡萄酒和白葡萄酒的酸度稍微不一樣，水果酒和白葡萄酒相同。

紅葡萄酒　5.5～6.5公克/公升

白葡萄酒/水果酒　6.0～8.0公克/公升

上面的數值是理想標準值，當你準備釀酒用的果汁或酒醪和上面數據有些微的差距時，不必介意也不必調整，如果酸度低於5公克/公升或高過9公克/公升，就需要增酸或降酸了。

舉例說明增酸的計算。你有20公升的紅葡萄汁，它的酸度是4.5公克/公升，你要將它增酸到6.5公克/公升，也就是說每公升要加入2公克的增酸劑，20公升總共需要20公升×2公克＝40公克的增酸劑。增酸劑要購買混合性增酸劑，是用酒石酸、蘋果酸、檸檬酸按比例調配而成的。

舉例說明減酸的計算。果汁或酒膠的酸度過高時，加入釀酒專用減酸劑即可將酸度中和。也可以使用食品級碳酸鈣。但要注意碳酸鈣的純度。你有20公升的梅子汁，它的酸度是14公克/公升，你要將它降到8公克/公升，每公升要降低6公克的酸。6公克×20公升＝120公克降酸劑。

另外，增酸劑或降酸劑在使用時要先磨碎，溶解在少量的果汁或酒醪中，再加入整桶的果汁或酒醪中，攪拌均勻。

增酸劑

降酸劑

測酸步驟

 →  →  →  → 

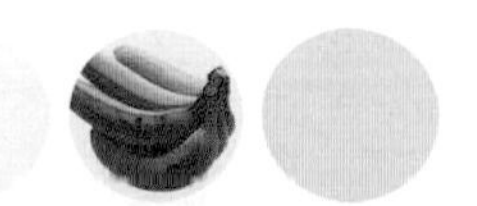

2.糖：檢測果汁或酒醪中的含糖量，比檢測酸的含量要容易多了，有兩個儀器很方便使用。

(1) fractometer　光譜測糖器

(2) Hydrometer　比重測糖器

家庭釀酒使用比重測糖器較方便，因酒醪發酵後產生酒精，酒精會影響光譜測糖器的精準度，而且比重測糖器售價也便宜多了。比重測糖器看起來像一個大溫度汁，用特製的直立塑膠管裝滿果汁或酒醪（若有果肉渣，須先過濾，只用澄清汁），把比重器放入直立管中，它漂浮在果汁中，眼睛平視直立管頂端，比重器的刻度即是糖的含量。

光譜測糖器

果汁或酒醪中糖的含量和成品酒的酒精度有直接關係，每2度的糖（2度＝1公升果汁中含有20公克的糖）經過發酵作用轉化成1度的酒精。理想的酒精度，紅葡萄酒是11～14%（v/v），白葡萄酒及水果酒是9～13%（v/v）；釀酒時先檢測果汁或酒醪的糖含量，大部份的水果都達不到20度以上的含糖量，補糖的計算和增酸的計算相同。

20公升果汁或酒醪的含糖度是18度，若想要提高到24度，每公升增加6度。

20公升×60公克＝1200公克＝1.2公斤

用蜂蜜釀酒，它的糖度太高了，可以用水稀釋，達到理想的糖度。

比重測糖器，測清水時是零度

3.酒精度：理論是這樣的，當你的果汁或酒醪含有24度的糖，經過完美的發酵作用，成品酒的酒精含量是12%（v/v）。請注意「完美發酵」這個條件，在發酵過程中，溫度、空氣、酵母菌都會影響酒精的含量。家庭釀酒使用比重測糖器得知果汁或酒醪中的糖含量，並不是非常精確，只是讓我們知道成品酒中的大概酒精含量。

如果你是個追根究底的人，一定要知道酒中的正確酒精含量，那麼就要多花點錢準備一套實驗室用蒸餾器，蒸餾出酒中的酒精再來計算它的正確濃度了。

4.二氧化硫：二氧化硫在釀酒中佔有很重要的功能，了解二氧化硫的作用，才能充份掌握釀酒的技術。

(1) 抑制作用：新鮮水果、蔬菜採收時，表皮帶有天然的酵母菌及雜菌、霉菌。利用水果蔬菜釀酒時，要加入適量的二氧化硫，抑制表皮上的各種野生菌增生繁殖的機會，避免污染。

(2) 抗氧化作用：空氣中的氧會溶解到果汁或酒醪中，溶氧中含有氧化酶，造成果汁或酒醪氧化褐變，二氧化硫會破壞氧化酶，有助於保持果香，減少新鮮度損失及褐變。

使用量：

為了抑制野生雜菌及防止氧化，果汁或酒醪需要維持一定的二氧化硫含量，發酵前要加入80～100ppm（1ppm＝百萬分之一）的含量，經過發酵後大部份被消耗或揮發，每次虹吸換瓶時，最好再加入30ppm的含量，最後裝瓶前再加入50ppm的含量，政府規定的安全含量是250ppm。

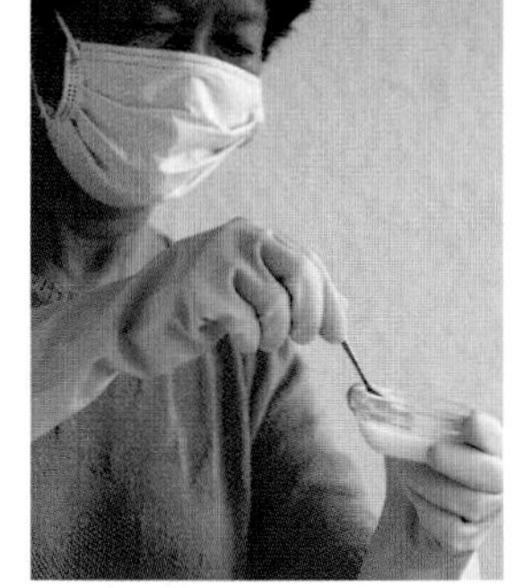

戴口罩、手套操作二氧化硫

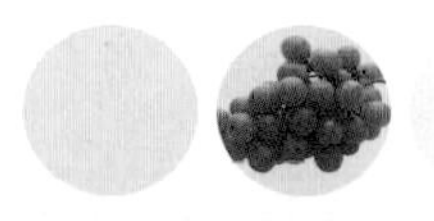

二氧化硫還有一個很重要的功能──殺菌，釀酒時使用的器具，洗乾淨後用高溫熱水殺菌，才能使用，若環境不容許或設備不足，不能有足夠的高溫熱水殺菌，則可以使用二氧化硫水溶液沖洗殺菌，4公升水中加入60公克二氧化硫粉末就是很好的殺菌劑，使用時避免皮膚和二氧化硫水溶液直接接觸，帶手套及活性碳口罩，以免吸入二氧化硫，有氣喘及二氧化硫過敏的人，千萬不要接觸二氧化硫。

5.溫度：果汁或酒醪發酵時，控制適當的溫度，是成功的第一要素，白葡萄酒和水果酒，最好不要超過20℃，紅葡萄酒是25℃，歐美的酒窖可以有很好的溫度控制，台灣的夏天幾乎都是30℃以上的大熱天，怎麼辦呢？就讓發酵桶泡冷水浴，將發酵桶放在另外1個裝滿了冷水的大桶中，溫度仍降不下來的話，在冷水中加入一些冰塊，發酵時溫度低點反而會提高酒的品質。

6.酵母：台灣大部份是使用冷凍乾燥的酵母菌，使用前要先將它活化，再加到發酵桶裡。方法如下：用1杯溫水（38℃左右），把需要的乾酵母菌溶在溫水裡，靜置10分鐘，若是冬天可以放置30分鐘，攪拌均勻再倒入發酵桶裡，20公升的果汁或酒醪需要5公克的酵母菌，發酵時能加入1茶匙的酵母營養劑更能助長酵母菌的生長繁殖，每一次釀酒時都使用新的酵母菌，不要為了節省酵母菌的花費（5公克酵母菌約30元）而使用回收的酵母菌，因小失大，損失了所有的水果原料。

冷凍乾燥酵母加入38℃溫水活化

酵母粉加水活化

7.澄清：澄清的定義是加入一種媒介物，借由它的作用，吸附或除去某種不好的氣味或物質，增加酒的香醇及穩定性，使酒液更清澈。

家庭釀酒，只要有耐心，多虹吸換瓶幾次，除去底部的沈澱，酒會自然澄清，但需要比較長的時間。現代人什麼都追求速度，凡事都要快，釀酒也不例外，使用澄清劑可以縮短等候的時間，早點喝到自己釀的酒。市售的澄清劑有很多品種及廠牌，基本上，紅葡萄酒用吉利丁粉，白葡萄及水果酒用皂土就可以達到很好的澄清效果。唯一要注意的是，要購買釀酒專用的吉利丁粉及皂土，下面用實例說明吉利丁粉及皂土的使用方法：

A.紅葡萄酒：20公升葡萄酒需要5公克吉利丁粉澄清。

方法：

（1）5公克吉利丁粉加入1/2杯冷水中，輕輕攪拌，放置10分鐘。

（2）隔水加熱，充份溶解成透明狀，但不要煮滾。

（3）倒入待澄清的酒液中，攪拌混合。

（4）靜置10天，虹吸換瓶。

吉利丁

已澄清的葡萄乾酒

※吉利丁含有正離子，吸附酒中的負離子單寧，而使酒液澄清，使用吉利丁粉時，酒中一定要含有足夠的單寧才會發揮作用，最常使用在紅葡萄酒的澄清。

B.水果酒：20公升水果酒需要2公克皂土。

方法：

(1) 2公克皂土加入1杯熱水（蒸餾水）。

(2) 用果汁機攪拌3分鐘（中速）。

(3) 加蓋放置24小時。

(4) 再用果汁機中速攪拌3分鐘。

(5) 慢慢的一邊攪動一邊倒入待澄清之酒液中，充份混合。

(6) 靜置10天，虹吸換瓶。

※皂土呈鹼性反應，它的膠體性質和本身帶有的負電荷，使它具有吸附力，白葡萄酒和水果酒含有很多蛋白質，帶有正電荷，容易和帶有負電荷的皂土結合，沈澱而分離。

8.陳年：日光和日光燈都會影響酒的品質，第一次發酵完成，虹吸到第二次發酵瓶後，一定要避免光線，放在陰暗的地方。如果用橡木桶做第二次發酵，不要裝得太滿，以免第二次發酵時，酒會滿溢出來。每一個月嘗一下橡木桶裡的酒，太濃的橡木味會蓋過葡萄酒的風味，反而有畫蛇添足之憾！

橡木桶

9.勾對：勾對是釀酒程序中很重要的一環，看起來很簡單，只是將兩種或兩種以上的葡萄酒進行混合，但要想得到最佳的效果卻非易事。

勾對的原則是將不同品種的葡萄，或不同產地的葡萄所釀製出來的葡萄酒，互相調配而形成獨特的風格。除了勾對後使酒變得更好，超越用來勾對的原酒，否則千萬別互相勾對，勾對的目的是為了讓基酒的某些特點加強或減弱，互補所長，達到複雜豐富的特色。勾對時紅葡萄酒和紅葡萄酒互相調配，白葡萄酒和白葡萄酒互相調配。先取小量的基酒，測出糖度、酸度、酒精度，以不同的比例互相混合，找出最理想的口味，再大量的勾對。

10.調味：利用糖、酸、高濃度酒精、果汁、香精、添加到釀好的葡萄酒或水果酒中，改變它的甜酸度、酒精度、香味、顏色等；謂之調味，即沒有一定的原則，憑個人的喜好決定。使用的添加物更達數百種，草藥、香草是歐洲人喜愛的添加物，苦艾酒（中國稱之為『味美思』）是最有名的代表，它是用酒精度18%左右的葡萄酒浸泡大約80種不同的植物，包括：花、草、樹皮、草根種子、樹葉等製成，是歐美人士喜愛的飯前酒。五加皮酒，竹葉青酒也都是調味酒。

香草

柳橙汁

11.裝瓶：新買的酒瓶，通常都附有灰塵，使用前一定要清洗，以免裝酒時受到污染。為了避免酒和空氣的接觸，利用虹吸方式注酒時，虹吸管一定要深到瓶底，讓酒由底部往上填滿酒瓶，千萬別把虹吸管插入酒瓶上端，讓酒像噴水似的注入酒瓶。

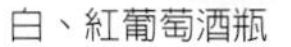

白、紅葡萄酒瓶　　香檳酒瓶

12.軟木塞：葡萄酒傳統使用軟木塞封口，但是市售的軟木塞品質良莠不齊，劣質的軟木塞容易發霉，污染了整瓶的酒。目前歐美酒莊逐漸開始採用樹脂合成的瓶塞，有軟木塞的優點但不會漏酒發霉。

傳統軟木塞及新式取代軟木塞之瓶塞

13.軟木塞機：軟木塞機有人工操作的一次一瓶，也有半自動到全自動的大型機器。

用槌子裝軟木塞敲入瓶口

用軟木塞封口

增加美觀性及封閉性之外套

如何用外套封住酒瓶：

1.將彩色的外套覆蓋住已用軟木塞封口的酒瓶

2.用吹風機加熱外套，使外套收縮將瓶口密封

3.已密封的酒瓶

The Keys to the Success of Winemaking

1. Acid: When there is not enough acid in wine, the wine will taste bitter, light and flavorless. If there is too much acid, it will be too dry to drink. Proper adjustment of the acid is extremely important.

Before making wine, the acid and sugar in the liquid and must have to be carefully measured. In the next paragraph we will discuss the measuring of the sugar. The most common way to measure the acid is to use an acid test kit. Remove the liquid or must carefully (if there are any solids, strain first to obtain a clear must). Drip the alkali solution drop by drop into the liquid that contains acid. Observe the changes. As the liquid or the must balances with the alkali, calculate the number of drops to determine the acid gravity of the must. Instructions will come with the acid test kit and should be carefully followed.

Color changes in the light colored liquid or must can be easily observed and it should be simple to tell if the acid and the alkali have balanced. Because of its dark purple color, with red wine it is a little trickier to observe the precise moment when the acid and the alkali are balanced. Here are two tips to smooth your way. One is to add distilled water to lighten the color. Because distilled water is neutral, it will not increase or decrease the acidity of the grape juice. Another method is to use a test kit. When you feel that the alkali drops are sufficient, drop a little liquid or must on the litmus paper and observe the color changes on the paper. If the litmus paper turns red, it means that it is still acidic and has not yet balanced; if the litmus paper turns blue, its means that it is too alkali, and you have probably added too much alkali.

What is the ideal acidity?

There is a slight difference between red and white wine. Fruit wines and white wines are the same.

Red wine 5.5 ~ 6.5 grams/liter

White wine/fruit wine 6.0 ~8.0 grams/liter

The digit above is the ideal alpha, when the liquid or the must you prepare is only slightly distant from the digit above, it is acceptable and it is not necessary to make adjustments.

If the acidity is below 5 grams/liter or more than 9 grams/liter, the acidity should be adjusted.

Here is an example of how to calculate acidity adjustments:

You have 20 liters of red grape juice whose acidity is 4.5 grams/liter and you want to increase the acid to 6.5 grams/liter. For every liter you must add 2 grams of acid blend.

20- liters of grape juice needs a total of 20 liters x 2 grams = 40 grams of acid blend.

Purchase acid blend which is a blend of identical proportions of tartaric acid, malic acid and citric acid.

Here is an example of how to calculate a reduction in acidity:

When the acid in liquid or the must is too high, add acid reducing formula or food-grade calcium carbonate to balance the acidity. Pay close attention to the purity of the calcium carbonate. If you have 20-liters of plum juice, the acidity is 14 grams/liter and the acidity should be reduced to 8 grams/liter. The acidity of every liter must be reduced 6 grams. 6 grams x 20 liters = 120 grams acid blend must be added.

Acid blend or acid reducing formula needs to be crushed first and dissolved in a small amount of liquid or must before adding to the fermentor. Once added, make sure to stir the liquid or must thoroughly.

2. Sugar: Measuring the sugar gravity in the liquid or must is much easier than measuring acidity. There are two types of equipment that can be used for this purpose.

A. Fractometer

B. Hydrometer

A hydrometer is more convenient in home winemaking because the must produces alcohol after fermentation. Alcohol will affect the accuracy of the fractometer. Also a hydrometer is much cheaper. It looks like a large thermometer with a specially-made rubber tube down the center. When filled with liquid (the must), it can measure its specific gravity. As the tube floats in the liquid read the scale inside the instrument at the level where the liquid contacts the glass.

The sugar content of the liquid or must has a direct relation to the alcohol content of the wine. Every 2 degrees of sugar gravity (2 degrees of sugar gravity = 1 liter liquid with 20 grams of sugar) after fermentation transforms into 1 degree of alcohol gravity. The ideal alcohol level of red wine is 11 ~ 14% alcohol by volume, for white wine and fruit wine it is 9 ~ 13% alcohol by volume. Measure the sugar content of the liquid and must, as most fruit will not exceed 20 degrees sugar content. The calculation for adding sugar is the same as that for increasing the acid.

20-liter liquid or must contains 18 degrees, and you want to increase it to 24 degrees, then each liter needs an increase of six degrees.

20 liters x 60 grams = 1200 grams = 1.2 kgs.

When using honey for winemaking, due to its high sugar content, dilute with water to reach the sugar gravity you desire.

3. Alcohol gravity: In theoretically perfectly efficient fermentation, when your liquid or must contains 24 degrees sugar gravity, the wine will have an alcohol content of 12% by volume. Please note that during fermentation, the temperature, air, and yeast affect the level of alcohol and reduce it below the ideal. Home winemakers use a hydrometer to measure the sugar gravity in the liquid and must, yet this measurement is not always accurate. It only provides an approximate estimate of the alcohol in the wine. To measure the exact level of alcohol in the wine, a scientific-quality distillation machine is required. Using this method , distill the alcohol from the wine, then measure its correct level.

4. Sulphite: Sulphite plays a very important role in winemaking.

a. Prevents microorganism growth: Natural yeasts, microorganisms, and bacteria live on the surface of the fresh fruit or vegetables. When using fruits and vegetables, add sulphite to prevent wild yeasts from multiplying and contaminating the wine.

b. Prevents oxidization: Oxygen in the air will dissolve into the liquid or must. The dissolved oxygen contains oxidization inhibitors and causes the liquid or must to oxidize and darken. Sulfur will destroy this and help to maintain the aroma of the fruit and reduce darkening and the loss of freshness.

Use amount:

In order to suppress wild bacteria and prevent oxidization, the liquid or the must needs to maintain a certain level of sulfur dioxide. Add 80~100ppm (1ppm = 1/1,000,000) prior to fermentation. After fermentation, most will be exhausted or evaporated, so add 30ppm each time when siphoning and racking to new bottles. Then add 50ppm when bottling. The government safety regulation is 250ppm.

Sulphite has another very important function: sterilization. Rinse all equipment before making wine, then sterilized under steaming hot water. If a good environment cannot be provided or sterilization equipment is not available (boiling water is not available) then sulfur dioxide solution can be used to sterilize. Sixty grams of sulphite powder for every 4 liters of water is an excellent sterilizer. When using, avoid direct contact with the skin. Wear gloves or activated carbon mask to prevent the skin from absorbing the sulfur dioxide. People who have asthma or allergies to sulfur dioxide should not handle sulfur dioxide solution at all.

5. Temperature: Control of the temperature during the liquid or must fermentation is the first key to success. The temperature of white wine and fruit wine is best kept below 20C, and red wine below 25C. The wine cellars in Europe and America utilize advanced temperature control systems. On a summer day in Taiwan the temperature is over 30C. What can be done? Give your fermentor a cold bath. Place the fermentor in another bucket filled with cold water. If the temperature still does not drop, add ice cubes to the cold water. Controlling the temperature during fermentation will increase the quality of the wine.

6. Winemaking yeast: In Taiwan, frozen dried yeast is most commonly used. Before use, revitalize the yeast and then pour into the fermentor. Follow these instructions: prepare a cup of lukewarm water (approximately 38C), dissolve the dried yeast in water and let sit for 10 minutes. Let sit for 30 minutes during winter months. Stir well, then pour into the fermentor. Twenty liters of must requires 5 grams of yeast. In order to help the yeast multiply and grow, add 1 teaspoonful of yeast nutrients during fermentation. Use a new pack of yeast every time you make wine and do not try to economize by using recycled yeast (5 grams of yeast is about NT$30).

7. Fining: The definition of fining is to add some kind of medium whose function is to absorb or remove unpleasant flavor or materials and to enhance the aroma and stability of the wine, so as to make the wine clear.
Home winemaking requires patience for the repeated siphoning and racking to new bottles to remove the sediment at the bottom. However long it takes, the wine will naturally be clear. Nowadays individuals pursue speed in all things and that includes winemaking. Using a fining agent can shorten the waiting period and enable early enjoyment of the wine. Fining agents available commercially come in many varieties and brands. Basically, red wine requires gelatin powder, while white wine and fruit wine use bentonite. Both provide excellent clearing effects. The only thing to keep in mind is that specialized gelatin powder and bentonite for winemaking must be used. The following are examples of how to use gelatin powder and bentonite:

A . Red wine: 20-liters of red wine requires 5 grams of gelatin powder to clarify.

Steps:

1. Dissolve 5 grams of gelatin powder in 1/2 cup of cold water, stir gently and let sit for 10 minutes.
2. Heat in double-boiler until dissolved completely or transparent, but do not bring to a boil.
3. Pour into the must in bottles and stir to combine.
4. Let sit for 10 days, siphon and rack to new bottles.

* Gelatin contains positive ions and absorbs the negatively charged tannins and clears the wine. If gelatin powder is used, the wine has to have enough tannins to enable its function. It is mostly used in clarifying red wine.

B. Fruit wine: 20-liters fruit wine needs 2 grams of bentonite.

Methods:

1. Add 2 grams of bentonite to 1 cup of hot water (distilled water).
2. Blend in blender for 3 minutes (medium speed).
3. Cover and sit for 24 hours.
4. Then blend over medium speed in blender for 3 minutes.
5. Pour into the must and stir at the same time until completely combined.
6. Let sit for 10 days, siphon and rack to new bottles.

* Bentonite gives an alkaline reaction, its gel properties and its negative charge give it the ability to absorb. White wine and fruit wine have lots of protein, which is positively charged, combines easily with the negatively charged bentonite and sediments and separates out.

8. Aging: Sun light and artificial light will both affect the quality of the wine. After the primary fermentation, siphon to bottles for the second fermentation and place them in a cool and dark location away from any light. If oak barrels are used for the secondary fermentation, do not fill the barrels to the top. This will prevent the wine from overflowing during fermentation. Taste the wine every month because too much oak flavor will mask the flavor of the wine. You may feel like an oak barrel is authentic and you are trying to get everything just right, but in reality it may turn out to be too much.

9. Mixing: Mixing wines is a very important step in the winemaking process. It appears simple, just mix two or more wines together. However, it is quite difficult to obtain the best effect. The principle of mixing is to combine two wines, from different types of grape, or from the same grape but from different places, to form a new wine with a unique character. Unless mixing wine improves the taste, it is better not to mix. The purpose of mixing is to strengthen or reduce certain features of the wine and to supplement the two wines in order to achieve some complex characteristic. Remove a small amount of wine, measure the sugar gravity, acidity, and alcohol, test different proportions and mix together to find the ideal combination, then next mix large amounts together.

10. Flavoring: Flavoring is the use of sugars, acids, alcohols, liquids, and fragrances in wines and adjusting factors such as sugar gravity, acidity, alcohol, fragrance, and color. There is no real principle, so experiment. Many different additives are possible. Herb medicines and herbs are preferred by Europeans. Vermouth is a well-known example. It is made by using grape wine with around 18% of alcohol with about 80 kinds of different plants, including flowers, grasses, tree bark, grass roots, seeds and leaves. It is a popular European aperitif. Chu Yeh Ching (bamboo leaf) wine are common flavored wines.

11. Bottling: New bottles are often full of dust. Rinse well before using to prevent contamination. To prevent contact between the wine and the air, use a siphon to suck up the wine. The siphon tube has to be long enough to reach the bottom of the bottle and let the bottle fill with wine from the bottom up. Do not insert the siphon tub in the opening of the bottle and let the wine spray into the bottle.

12. Cork: According to wine making tradition, soft corks of corkwood are used. However, commercially available soft corks are of inconsistent quality. Bad corks can quickly ruin an entire bottle of wine. Recently European wine cellars have started to shift to corks made of resins or plastics, which have the advantages of soft corks, but do not have the negatives of leaking or contamination.

13. Cork Machine: There are various kinds, ranging from almost completely manual to partially and fully automated. Some require one person to operate the machine while, others are large fully-automated factory-like machines.

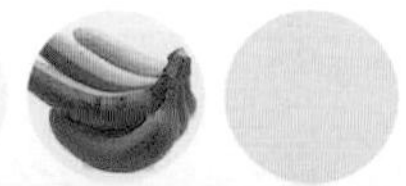

品酒

依照正確的品酒步驟，更能優雅從容地享受美酒的滋味。

1. 準備一個標準型葡萄酒杯，品酒時手握杯腳以免手掌端著酒杯，會使掌心的溫度增加了酒的溫度。
2. 倒酒：緩緩地倒入1/3杯的酒，有充份的空間讓酒的香氣上升。
3. 看：舉起酒杯，面向亮光處，觀察酒的顏色是否清澈、明亮。
4. 聞：輕輕旋轉酒杯，將鼻子埋在酒杯中間，深深吸一口氣。首先，你要找出是否有葡萄的果香，然後找出酒香，葡萄及其它水果在釀製中經過橡木桶陳年，或其他不同的釀製程序，會產生不同的酒香。酒香加果香被歐美人仕稱為「NOSE」，不是指鼻子，而是品酒者的專用名詞，指酒特有的香醇之氣。
5. 含：千萬別太急，慢慢地吸一點酒到口內，讓酒在口腔內滾動，經過舌尖及兩頰，此時你的味蕾及味覺會告訴你很多資料，許多人先天具有這種品酒才能，也許你就是其中之一。
6. 嚥：專業品酒師很少將酒嚥下肚，此時他們會將酒吐出來。我們品酒更想喝酒，此大好良機當然一口嚥下，經過舌根流入胃中，「哈」呼一口大氣，酒若有苦澀味，此時就逃不過你的品嘗。舌頭是人體上最奇妙的器官之一，它的前端會研判出甜味，兩旁偵測出酸味，舌根則是苦澀味，能使人類避免吞下有毒的食物。
7. 停：回想前面動作的總結，口中留下的餘味是否很快就沒了？還是仍帶有耐人的回味？
8. 談：品嘗酒是最好的談話資料，不牽涉政治、性別、階級，藉著彼此交換品嘗酒的結果，找出每個人不同的味覺及感覺。

Tasting

1. Prepare a standard size wine glass, hold the stem to prevent your body heat from heating the wine.
2. Pouring: Pour the wine gently into the glass until it is about 1/3 full, the remaining space will permit the aroma of the wine to rise.
3. Observing: Hold your glass up to the light and observe the color of the wine. Is it clear and brigjt?
4. Smelling: Turn the glass gently, bury your nose in the glass and breathe in deeply.First, you have to determine whether there is the aroma of grape, and then you must locate the bouquet of the wine. Grapes and other fruits will produce different aroma and bouquet during the winemaking and aging process.
5. Hold the wine in your mouth: Without rushing, sip a little wine into your mouth slowly and let the wine roll around your mouth, over the tongue and around the cheeks.
6. Swallowing: The wine goes through the tip of the tongue and flows down into your stomach. Exhale with a 'ha!" If the wine has a dry flavor, it cannot escape your perception. The tongue is one of the most particular organs of the body. The front tastes sweetness, but the sides taste sour, and the rear finds dry flavors. It protects human beings from swallowing anything poisonous.
7. Stopping: Is there any remaining taste in your mouth? Is it gone already? Or does some pleasant flavor still linger in the memory?
8. Talking: Wine tasting makes for fine conversation, since it can be done regardless of sex or politics or social class, it can stimulate conversation and interpersonal exchange.

釀酒Q&A Wine Making Question & Answers

問題1：酵母菌活化後加入發酵桶48小時仍沒有發酵現象。

原因 1.酵母已經過期 2.加了過量（計算錯誤）二氧化硫 3.酒醪太酸 4.溫度過低

解決 1.重新加1包酵母 2.降低二氧化硫含量（參閱問題10） 3.降低酸度（技術篇：酸） 4.用電毯圍裹或放在盛滿熱水的大盆裡升溫

Question 1 : There are no fermentation phenomena 48 hours after the vitalized yeast has been added to the fermentor.

Reason

1. The yeast is expired
2. Added too much sulphite dioxide due to wrong calculation
3. The must is too sour
4. The temperature is too low

Solution

1. Add another pack of yeast
2. Lower the sulphite dioxide level (see Question 10)
3. Decrease the acidity (Technique: Acid)
4. Wrap in an electric blanket or remove to a basin filled with hot water to increase the temperature.

問題2：發酵作用微弱，持續長時間發酵作用仍不停止。

原因 1.酒醪溫度太低 2.酵母缺乏營養劑

解決 1.用電毯或熱水浴增溫 2.添加1小匙酵母營養劑

Question 2 : The fermentation is very weak, it goes on for too long without stopping.

Reason

1. The temperature of the must is too low
2. The yeast does not have enough nutrients

Solution

1. Use an electric blanket or hot water to warm the fermentor
2. Add a small spoonful of yeast nutrients

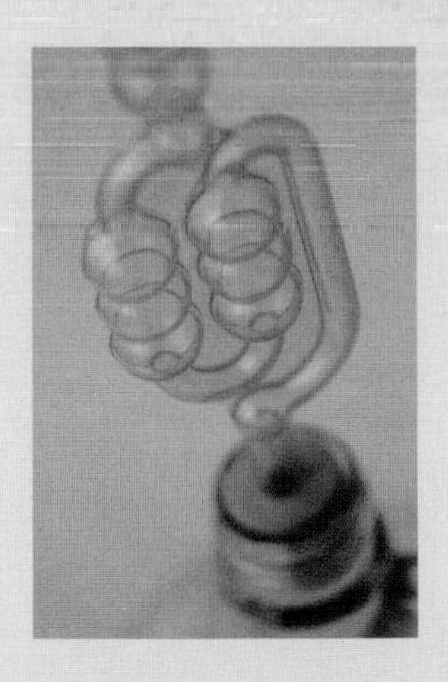

問題3：酒太酸了。

原因 1.水果不夠成熟 2.測酸時計算錯誤 3.降酸時計算錯誤

解決 酒已釀成，太酸的話只有補糖、提高甜度來中和酸味。

Question 3 : The wine is sour

Reason

1. The fruit is not ripe enough
2. Wrong calculation while measuring acidity
3. Wrong calculation while reducing the acid

Solution

Since the wine has already been bottled, if it is too sour, the only thing is too add sugar and to increase sugar gravity, so as to balance the acid.

問題4：酒有苦味。

原因 1.酒已經氧化了 2.酸度太低

解決 1.酒有苦味，表示氧化得太嚴重了，沒什麼補救之道，警惕自己下次釀酒時不要犯同樣的錯誤。2.增酸。

Question 4 : The wine is bitter

Reason

1. Wine has become oxidized
2. The acidity is too low

Solution

1. If wine is very bitter, it means that the oxidization is very serious. There is not much you can do now, just chalk it up to experience, and on your next attempt , avoid making the same mistakes.
2. Increase the acidity

問題5：酒有臭蛋味（H_2S氫化硫作怪）。

原因 1.發酵時溫度太高，發酵太快速。2.第二次發酵時間太久，酒醪和沈澱物接觸過久。

解決 用力攪拌酒醪，同時加入50ppm二氧化硫，用銅網過濾，可以除去輕微的H_2S。

Question 5 : There is a strong rotten egg flavor in the wine (caused by H2S hydrogen sulfide)

Reason

1. The temperature is too high and the fermentation is going too fast during fermentation
2. The secondary fermentation is too long, and the wine has been in contact with the sediment too long.

Solution

Stir the must vigorously, at the same time, add 50ppm sulfur dioxide, and strain with a copper strainer to remove any H_2S.

問題6：裝瓶熟成時，瓶底有沈澱。

原因 1.虹吸時不小心，將沈澱物吸入瓶中。2.裝瓶時沒有加安定劑，增加酒的穩定性。

解決 用澄清劑去除沈澱，再用過濾機過濾。

Question 6 : There is sediment when bottling for aging

Reason

1. Siphoning was done carelessly and sediment has been sucked into the bottle.
2. Did not add stabilizer when bottling to increase the stability of the wine.

Solution

Add clearing agent to remove the sediment, then strain.

問題7：裝瓶熟成時，酒變混濁。

原因 被雜菌感染。

解決 嘗一點混濁酒，若仍能喝沒有異味，用微過濾機（0.4Micron）除去雜菌，再添加50ppm二氧化硫。

Question 7 :The wine becomes cloudy when bottling for aging.

Reason

Contaminated by bacteria.

Solution

Taste the wine a little, if it is still drinkable and has no unusual flavors, use a filtration machine (0.4Micron) to remove the bacteria, then add 50ppm sulfur dioxide.

問題8：第二次發酵後，使用澄清劑，酒液仍然混濁。

原因 1.單寧含量太低 2.果膠酵素加的量不夠。3.雜菌感染。

解決 1. 20公升酒，加入2小匙單寧，靜置7天，再用皂土澄清，加入30ppm的二氧化硫。2.添加2小匙果膠酵素，待酒醪澄清時，虹吸換瓶。3.參閱問題7。

Question 8 : After adding the clearing agent, the wine is still cloudy after the secondary fermentation.

Reason

1. The amount of tannins is too low.
2. There is not enough pectin enzyme added.
3. Contaminated by bacteria.

Solution

1. Add 2t of tannin to every 20-liter wine. Let sit for 7 days, then clear with bentonite and add 30ppm sulfur dioxide.
2. Add 2t of pectin enzyme and wait until the must is clear, siphon and rack to new bottles.
3. Refer to Question 7.

9：酒聞起來有醋味。

原因 清潔殺菌不徹底，被醋酸菌感染。

解決 若是極輕微的醋味，趕快加入50ppm二氧化硫，醋酸菌對二氧化硫沒有抵抗力。若醋酸味太重，只能祝你好運，希望感染的是好醋酸菌，還能有一大桶水果醋，如果是壞醋酸菌只有倒掉了。

Question 9 : The wine smells like vinegar.

Reason

The cleaning and sterilization has not been well done, and the wine is contaminated by Acetobacillus.

Solution

If wine has a slightly vinegar flavor, add 50ppm sulphite immediately, Acetobacillus has no immunity to sulfur dioxide. If the vinegar flavor is too heavy, hopefully the Acetobacillus is a good one. If so, then you have a large barrel of fruit vinegar, if not, then you just have to toss it out.

問題10：計算錯誤，加了過量的二氧化硫。

原因 1.添加了過量的二氧化硫

解決 1.酒醪擴大1倍，稀釋過量的二氧化硫，這是最理想的解決方法。2.酒醪加熱到33℃，離火，用力攪拌，讓過量的二氧化硫揮發，直到酒醪冷卻。

Question 10 : Adding too much sulphite due to wrong calculation.

Reason

Added too much campden tablets

Solution

1. The ideal solution is to add double the amount of must to dilute the sulphite.
2. Heat the must up to 33℃, and remove from heat. Stir the must vigorously until cooled to permit the excess sulfur dioxide to evaporate.

TASTER001

冰砂大全 —112道最流行的冰砂

蔣馥安 著　定價230元　特價199元

ISBN=957-0309-36-9　CIP=427.46　特24開全彩

■水果冰砂、咖啡紅茶冰砂、調味冰砂、水果花草茶冰砂、果粒冰砂、花式冰砂和星座冰砂。

■本書精華：開店創業的最佳參考：提供時下最流行的冰砂配方、開店須知及工具材料供應商，輕鬆完成開店的夢想。

TASTER002

百變紅茶 —112道最受歡迎的紅茶．奶茶

蔣馥安 著　定價230元

ISBN=957-0309-37-7　CIP=427.41　特24開全彩

■以最簡易的製作方法、正確的泡茶知識，教你沖泡時下最受歡迎的優質冰熱紅茶、綠茶及奶茶、養生茶。

■開店創業的最佳參考依據：提供時下最受歡迎的紅茶奶茶配方、開店須知及工具材料供應商，幫助你完成開店的夢想。

TASTER003

清瘦蔬果汁 —112道變瘦變漂亮的蔬果汁

蔣馥安 著　特價169元

ISBN=957-0309-39-3　CIP=427.46　特24開全彩

■本書精華：暢飲蔬果汁的第一選擇：千變萬化的口味、詳盡的做法說明，不但可以品嘗新鮮衛生的蔬果汁，還能擁有漂亮膚質及勻稱好身材。開店創業的最佳參考書：提供時下最受歡迎的蔬果汁配方、挑選蔬果的祕訣、開店須知及工具材料供應商，讓你完成開店的夢想。包含112變瘦變漂亮的蔬果汁：60道果汁、28道蔬果汁、24道星座果汁。

TASTER004

咖啡經典 —113道不可錯過的冰熱咖啡

蔣馥安 著 定價280元

ISBN=957-0309-47-4　CIP=427.42　特24開全彩

■假使每天只喝1、2杯咖啡，當然，你要選擇最好、最特別的咖啡。

■教導讀者正確的運用各種咖啡器皿煮出好喝咖啡。而如果你想開一家咖啡店，書中還有開店須知，單品熱咖啡課程，教你選咖啡豆、烘焙咖啡、調配咖啡以及保存咖啡。

TASTER005

瘦身美人茶 ——90道超強效減脂茶

洪依蘭 著　定價230元，特價199元

ISBN=957-0309-70-9　CIP=418.914　特24開全彩

■蒐集最流行、最有效、最受歡迎的減脂茶和果汁，包括超強效瘦身茶、單味瘦身茶、紅綠茶、花草茶、中藥草茶等。

■90道茶譜各有特色，可以去水利濕、排除油脂、調節新陳代謝，且適量喝絕不傷身，建議讀者可以多多嘗試多換口味，即便無法立即減重也絕對有助消化。

TASTER006

養生下午茶 ——70道美容瘦身和調養的飲料和點心

洪偉峻 著　定價230元

ISBN=957-0309-73-3　CIP=418.914　特24開全彩

■50道沖煮簡便的養生茶飲與20道美味的養生茶點，讓讀者輕鬆利用簡單的飲品，在短短的下午茶時段，輕鬆達到保養身體的功效，為健康把關。

■茶飲針對現代人最著重的潤膚美白、減脂瘦身、強健養身與疾病調養四方面。

TASTER007

花茶物語 ——109道單品複方調味花草茶

金一鳴著　定價230元

ISBN=957-0309-85-7　特24開全彩

■本書以花草茶為主題製作出109道單品、複方花草茶，以及奶茶、咖啡、調酒、花草茶與冰涼的蔬果飲品，還有相卓料理教學與點心製作等單元。

■讓喜愛花草茶的朋友可以藉由飲品與可口的點心一起親近芳香的花草世界。

TASTER008

上班族精力茶 ——減壓調養增加活力的嚴選好茶

楊錦華 著　特價199元

ISBN=957-0309-73-3　CIP=418.914　特24開全彩

■針對工作忙碌的上班族精心設計53道優質茶方，讓讀者輕輕鬆鬆擁有健康的身體，除卻一身疲勞與壓力。

■每道茶飲各有特色，擁有舒緩全身痠痛、調理文明病、增強免疫力……等療效，讓沒有時間的上班族藉由飲茶改善身體，維護健康。

TASTER009

纖瘦醋 ——瘦身健康醋

徐因 著　定價199元

ISBN＝957-0309-91-1 CIP＝427.1　特24開全彩

■本書教導讀者利用常見水果蔬菜來製醋，包括蕃茄醋、蘋果醋等22道，並利用各式蔬果醋加上果汁、花茶等，設計出好喝且富變化的健康醋飲及醋料理共70道。

COOK50043

釀一瓶自己的酒

——氣泡酒、水果酒、乾果酒

作者 錢　薇
版型設計 許淑君
文字編輯 張立萱
美術編輯 鄧宜琨
企畫統籌 李　橘
發行人 莫少閒
出版者 朱雀文化事業有限公司
地址 北市基隆路二段13-1號3樓
電話 02-2345-3868
傳真 02-2345-3828
劃撥帳號 19234566 朱雀文化事業有限公司
e-mail redbook@ms26.hinet.net
網址 http:// redbook.com.tw
總經銷 展智文化事業股份有限公司
ISBN 957-0309-96-2
初版一刷 2003.09

定價 320元
出版登記 北市業字第1403號

國家圖書館出版品預行編目資料

釀一瓶自己的酒：氣泡酒、水果酒、乾果酒/
錢薇著. -- 初版 --臺北市：朱雀文化，
2003[民92]
面； 公分. --(COOK50；43)

ISBN 957-0309-96-2(平裝)

1. 酒 - 製造
463.81 92012345